Biocides in Synthetic Materials

2nd International Conference 28 - 29 September 2010, Berlin, Germany 2010

Organised by

SMITHERS
iSmithers

Berlin, Germany
28-29 September 2010

ISBN: 978-1-84735-554-6

Contents

Session 3: Functional Modification of Synthetic Materials to Achieve Effects in Textiles and Medical Devices

Paper 11   **Relating function to claims for antimicrobial properties in textile applications**
Pete Askew, IMSL, UK

Paper 12   **Antimicrobial textile effects**
Dr Heinz Katzenmeier, Sanitized AG, Switzerland

Paper 13   **A durable, light and heat stable and easy-to-use antimicrobial product for textiles and non-wovens**
Dr Tirthankar Ghosh, The Dow Chemical Company, USA

Paper 14   **The control of odour in sports and fashion wear**
Mike Sweet, Polygiene AB, Sweden

Paper 15   **Applications for silver in textiles and medical devices**
Peter Steinrücke, BioGate, Germany   *+++ paper unavailable +++*

Paper 16   **Antimicrobial technologies for contact lenses and contact lens cases**
Dr Manal M Gabriel, Ciba Vision Corporation, USA

Session 4: Functional Modification of Synthetic Materials to Achieve Effects on Non-Porous Surfaces

Paper 17   **Considerations in the design and development of an antimicrobial delivery system**
Paul Lawrence and Debbie Stephenson, DuPont Teijin Films, UK

Paper 18   **Factors affecting biocide compatibility and performance in PVC during processing and product lifetime**
Dean Nichols, Akcros Chemicals Ltd, UK

Paper 19   **The use of reactive silane chemistries to provide durable, non-leaching antimicrobial surfaces**
Bob Monticello, Aegis Environments, USA

<u>KEYNOTE PRESENTATION</u>

BIOCIDES IN SYNTHETIC MATERIALS: WHAT DO THEY NEED TO DO?

Dr Ina Stephan
BAM
Unter den Eichen 87, 12205 Berlin, Germany
Tel: 0049 30 8104 1412 Fax: 0049 30 8104 1417 email: ina.stephan@bam.de

BIOGRAPHICAL NOTE

Dr. Ina Stephan has a Ph.D in Wood Science and Technology from the Univ. of Hamburg, Germany and is head of the working group on the resistance of materials to microorganisms at BAM (Federal Institute for Materials Research and Testing of Germany). Her main interests are in the testing of materials against fungi, algae and bacteria in a meaningful manner and the international harmonization of test standards for biodeterioration.

ABSTRACT

Biocides are used in a wide range of materials of both synthetic as well as natural origin. Natural materials such as wood or wool often have very limited natural durability and environmental processes degrade and recycle the material by physical, chemical and biological means. However, mankind often aims to prolong the service life of these products especially when value has been added by using these materials e.g. for building purposes or clothing. The use of biocides is often welcomed when other means to protect a natural material (e.g. keep it dry, out of reach for insects) cannot be achieved.

Biocides are used in a wide range of materials of both synthetic as well as natural origin. Natural materials such as wood or wool often have very limited natural durability and environmental processes degrade and recycle the material by physical, chemical and biological means. However, mankind often aims to prolong the service life of these products especially when value has been added by using these materials e.g. for building purposes or clothing. The use of biocides is often welcomed when other means to protect a natural material (e.g. keep it dry, out of reach for insects) cannot be achieved.

Synthetic materials can also be susceptible to biological attack (see Table 1):

Here a preservative is often used for the same reasons as for natural fibres: the protection of the material itself and therefore its function in a certain situation.

However, many synthetic materials are not bio-degradable (see Table 1), but nevertheless biocides are incorporated to protect them in service. In the case of mould fungi an indirect attack on materials can occur when fungi or, in a broader sense, a biofilm colonises a surface metabolising e.g. dust, grease, perspiration, and other contaminants that were deposited on a surface during manufacture or have accumulated during service. Microbiological growth on a soiled surface can cause damage to the underlying material, even though that material may be resistant to direct attack. It can harm the synthetic material in different ways:

a) Metabolic waste products (i.e. organic acids excreted by fungi) can stain or degrade plastics.
b) fungal mycelium consists of filamentous hyphae which can form undesirable electrical conducting paths across insulating materials, for example, or may adversely affect the electrical characteristics of critically adjusted electronic circuits.
c) The biofilm can adversely affect light transmission through an optical system, block delicate moving parts, or change non-wetting surfaces to wetting surfaces with resulting loss in performance.

d)   The presence of the biomass itself (e.g. algae, fungal spores) defaces the surfaces on which it is growing disturbing its aesthetics.

The material is protected by a biocide as long as the biocide stays active, is bio-available and is not leached or evaporated from the material.

Table 1: Excerpt of MIL-STD-810G, METHOD 508.6 ANNEX B

Table 508.6B-I. Fungi susceptibility of materials. Group I – Fungus-inert materials (Fungus-inert in all modified states and grades)	
Acrylics	**5** Polyamide
Acrylonitrile-styrene	Polycarbonate
Acrylonitrile-vinyl-chloride copolymer	Polyester-glass fiber laminates
Asbestos	Polyethylene, high density (above 0.940)
Chlorinated polyester	Polyethylene terephthalate
Flourinated ethylenepropylene copolymer (FEP)	Polyimide
Mica	Polymonochlorotrifluoroethylene
Plastic laminates: Silicone-glass fiber	Polyporpylene
Phenolic-nylon fiber	Polystyrene
Diallyl phthalate	Polysulfone
Polyacrylonitrile	Polytetrafluoroethylene
	Polyvinylidene Chloride
	Silicone resin
	Siloxane-polyolefin polymer
	Siloxane polystyrene
Group II – Fungus nutrient materials **(May require treatment to attain fungus resistance)**	
ABS (acrylonitrile-butadiene-styrene)	Polydichlorostyrene
Acetal resins	Polyethylene, low & medium density (0.940 and below)
Cellulose acetate	Polymethyl methacrylate
Cellulose acetate butyrate	Polyurethane (ester types are particularly susceptible)
Epoxy-glass fiber laminates	Polyricinoleates
Epoxy-resin	Polyvinyl chloride
Lubricants	Polyvinyl chloride-acetate
Melamine-formaldehyde	Polyvinyl fluoride
Organic polysulphides	Rubber, synthetic
Phenol-formaldehyde	Urea-formaldehyde

Biocides can act in different ways and the purpose of using a biocide should be clear:

a) Preservation:
The aim of preservation is to prevent spoilage, decay or conglomeration of biomass which is detrimental to the functionality of an item or material. The intention of preservatives is not to transfer their effects to other materials or the environment, but to protect the material itself.

b) Disinfection:
The aim of disinfection is to kill at least a certain fraction of living cells in a short time no matter whether they spoil or decay a material. Also here, the intention is not to transfer their effects to other materials or the environment.
E.g. a surface once disinfected can become contaminated again immediately afterwards with a bacterial population.

c) Reduction via a treated article:
a reduction of micro-organisms on the surface of an item is intended so as to reduce transfer to any other surface. The target organisms have not necessarily an effect on the treated item. Other than in disinfection the kill rate is not > 99 % and can be much lower, but the treated article shall have a long lasting effect.

Presently there is a debate whether a treated article is already effective when it lowers the growth rate compared to an untreated article (growth inhibition), or whether it needs to actually reduce the initial number of microorganisms it gets into contact with.

Biocides are needed to solve problems such as those mentioned above in a large number of different environments. Sometimes immediate contact with mammals is a critical factor, sometimes a closed industrial process needs to be protected. Essential for the application is to define the problem, the aim/intention and the claim that is finally being made for the biocidal product. It is likely that also the benefit of a biocidal product will be questioned at some point.

Reflecting these differentiations The European Biocidal Product directive (BPD) has 4 main groups (disinfectants, preservatives, pest control and other biocidal products) containig 23 product types (3).

To structure the complex issues of product claims regarding efficacy of a biocide, a tiered approach is, in general, accepted by the involved authorities. In the Biocidal Product directive this is reflected by distinguishing between the evaluation of a biocidal active substance and a biocidal product. A biocidal active substance is often required to show efficacy in a model compound of the product type it claims to protect (e.g. synthetic coatings) against the type of organisms it claims to be active against. Its performance regarding leachability, UV-stability etc in this model matrix is not being questioned at this stage. These questions will arise when a biocidal product is to b addressed where weathering and/or other environmental conditions that occur during the use of a product can be of importance.

This tiered approach can be summarized very briefly:

A) Tier 1: Does the biocide work in principle in its appropriate model matrix?
B) Tier 2: Does the biocide show efficacy under the real life conditions? For how long?
C) Tier 3: Does the use of the biocide yield a benefit in practice?

When moving up through tier 1 to tier 2 and to tier 3 the more tailored to the field of application envisaged a test design has to be. While in tier 1 some existing standards are basically suitable when the biocide is tested in a relevant matrix with defined organisms and under reproducible conditions (which are normally only to be found in a laboratory) testing for tier 2 is often more complex and specific standards often do not exist. This is even more so for tier 3 tests where there may be a need for weathering cycles, wind tunnel tests, cleaning regimes etc. to evaluate whether the biocide maintains its efficacy when exposed to the environmental stresses it is likely to face in service. Similarly, soiling and the influence of other micro-organisms can be of more significance. When aging is performed in the field or under in use conditions the reproducibility can become a difficult issue, as the aging factors such as e.g. evaporation and soiling are difficult to reproduce and can influence the results.

Despite this, the questions posed at Tier 3 can be relatively easy to answer when only the protection of the material is being considered. However, when health benefits for the consumer are being implied, the questions often require complex studies to deliver an answer. After all, the vast majority of microorganisms are not by themselves dangerous. The healthy human body is able to cope with a number of potentially "dangerous" microorganisms. But, while this may be true for many domestic environments and for a healthy individual, it may not be the case in, for example, hospitals and nursing homes and for immuno-compromised individuals. Here, smaller numbers of organisms might have a more significant effect. One of the tasks for the tier 3 level of testing is to to explore the benefit that a treatment might deliver for example in reducing the size of populations on environmental surfaces and whether this has an impact on hospital acquired infections.

Thus the type of tests required to explore biocidal efficacy will depend on both the claim that is being made and the level (tier) at which it will be applied. Several examples shall be given to demonstrate how varied claims and approaches for generating data to prove such claims can be.

Example 1:
In-Can-preservation for containers holding synthetic materials e.g. spray foams, paints, glues, silicone. Many water-based products are susceptible to bacterial or fungal growth. The growth can be sustained by the synthetic material itself or by soiling of the material during the production process.

Does the product support bacterial growth under the conditions of the lab test when there is no preservative added? If so, you might want to add a biocide in order to stop any damage or biological deterioration of your product – you want to *preserve* your material.

Does the preservative used demonstrate a biocidal effect, does it reduce growth rates or does it limit metabolism compared to an unpreserved material?

Claim: The biocide preserves the water-based synthetic material in-can; no spoilage will occur due to bacterial growth.

This is a classic claim for in-can preservatives of all kinds. The rational behind it is that materials, like adhesives, paints, foam-fillers etc can potentially be colonised by bacteria. These bacteria can metabolise certain components of the product and this biochemical change can lead to bad smells, change of colour, loss of viscosity, changes in pH and other effects that negatively influence the performance of the product. A preservative is added to preserve the state of the product so it does not change. It is assumed that any bacterial metabolism in the product is detrimental to it. Metabolism of bacteria can be detected in different ways, e.g. CO2 emission, O_2 depletion or, more commonly, by an increase in the number of viable cells present. For example, to measure the number of viable cells in a product an aliquot is usually taken from the material (e.g. paint) in the can and transferred onto a nutrient media, in some cases after dilution in a neutraliser for any preservative that may be present and to ensure the population can be measured. Once on the medium, the cells can proliferate and form visible colonies. Counting these colonies provides an estimate of the number of viable cells present in the system (as colony forming units – CFU).

However, many microorganisms are able to form dormant cells or spores to survive unfavourable environmental conditions. These resting cells do not proliferate and show no significant metabolic activity until they find a suitable environment. It is therefore possible that vital and active cells, being exposed to an unfavourable environment e.g. a synthetic paint containing solvent or even a preservative, are transformed into dormancy. Only when a sample of the material is taken and is spread onto a nutrient medium do the cells start to grow and to build new colonies. This underlines that the appearance of colony forming units on a nutrient media is not necessarily sufficient evidence that growth is occurring in a material.

Therefore, even in a tier 1 test, in order to prove growth, and with this also metabolism and changes in a material, you will have to observed the response of a population over time. If no growth can be shown and no metabolism can be proven then the need for a preservative is questionable and should be reconsidered for the sake of the protection of consumers and the environmental as well as for economic reasons.

In a tier 2 test other parameters have to be considered to test efficacy of the preservative. Most commonly aging procedures are being applied to help establish a shelf life e.g. frost, high temperatures, condensation. Useful field tests are relatively rare in such systems, but can be of value in applications that are difficult to simulate in the laboratory e.g. tinting machines used for paint.

Example 2:
Plastic benches in the changing room of a public swimming pool.

Plastic benches attract a multitude of soiling material during use that can act as nutrients for micro-organisms: from body lotion, skin cells, to sticky food. This nutrient film can easily support a growing biofilm when water is present. Mould fungi, e.g. Aspergillus sp., might damage the material by emitting organic acids or spores might permanently discolour the surface by resting in small pores and fine cracks of the plastic. In order to prevent these effects a fungicide can be incorporated into the plastic.

A solid matrix is in some ways more of a challenge to protect by incorporating a biocide than a liquid matrix. The keyword here is "bioavailability". The biocide has to get into contact with the microorganism to act. If the biocide molecule is "stuck" in the matrix it cannot act (unless the molecule is on the very surface of the plastic). To make the biocide available as a constant film on the surface it can either be applied as such or diffusion must occur from the inside of the material to the outside.

Claim: The fungicide preserves the solid synthetic material; no disfigurement of the surface due to mould growth will occur

In tier 1 of a testing regime the basic principle needs to be proven. In this example, that the chosen preservative will preserve a plastic material against mould growth. As in example 1 it is important to show in the test that an untreated material supports growth that leads to damage. Commonly a nutrient media is applied in these tests to mimic the soiling of the surface in practice. If mould growth is stopped on the biocide treated surface of the samples the claim for tier 1 would be met.

A tier 2 test might include aging of the material by bringing it into contact with acidic or alkaline substances or by leaching in in water before the biological test is performed.

If the plastic bench were outside, UV-treatments might be appropriate to test the stability of the biocide under this condition.

Example 3:
A lack of hygiene is often the cause for spreading of diseases from common colds to severe outbreaks of salmonella infections. For some years now articles are on the market claiming to prevent contact infections by killing any bacteria that settle on the surface of items as e.g. light switches, toilet seats, wash cloths or refrigerator shelves. Silver-ions, often referred to as "nano-silver", are applied in this context. The primary purpose of biocides in treated articles is **not** to preserve the material and its function (e.g. stability in load carrying constructions), but to hinder growth of bacteria or to reduce their number on the surface of a treated article. The intention is often to show a positive effect concerning hygiene.

Claim: The treated article shows hygienic effects and limits transmission of bacteria or diseases caused by these.

Basic bactericidal activity in this claim could be demonstrated with tests like ISO 22196 (2007): Plastics – Measurement of antibacterial activity on plastic surfaces" (4). The test is designed such that a small area of the synthetic material under test is held in intimate contact with a defined number of bacteria, suspended in a dilute solution of nutrients, for 24 hours at 35°C. During the test a cover (e.g. polyethylene) prevents the surface from drying and ensures intimate contact. After 24 hours the residual bacteria are washed off of the surface using a neutraliser and an aliquot of this solution is transferred onto nutrient media as described for example 1. Several outcomes of this test are possible. The surface could have acted:

A, as biocidal; the initial number of bacteria that were transferred to the surface has been reduced
B, as biostatic or rather causing "static-growth"; compared to the initial number of bacteria and provided that there is growth on the untreated, biocide-free surface
C, as limiting growth; compared to the growth on the untreated, biocide-free surfaces tested under the same conditions.

Other than for a preservative, the biocide-free surfaces do not necessarily have to show growth. An effect can already be claimed when the die-back of bacterial cells is faster on the treated surface than on the untreated surface.

Considering the above claim for this example all test results (A, B and C) could support it. Only if there was no growth on the untreated surface, or the same or more growth on the treated surface as on the untreated surface the test would the claim not substantiate at tier 1.

In tier 2 the field of use envisaged has to be taken more into account to justify a claim. As before, performance during use has to be considered. For example, a cutting board, claiming to have anti-microbial properties, is likely to be washed many times in a dish-washer. Is the anti-microbial property persistent?

Also, in practice, frequent drying and wetting of the surface is much more likely than 24 hours of water contact. How effectively is the biocide getting into contact and reacting with the bacterial cells under these circumstances? Are the bacteria on the untreated surface dying "a natural death" caused by dissication?

It is unlikely that a cutting board used in a private household will be radiated with UV light, but in industrial applications it might such exposure may well be employed to attempt to reduce microbial populations.

Testing for tier 3 needs to be carefully tailored to deliver an answer. Is there an end-user scenario in which there is a health benefit in applying the biocide-treated article?

Example 4
Traffic on roads and streets is regulated by signs. These signs contain important information for road safety and need to readable under all circumstances. Algal growth can potentially disfigure the surface of such signs and obscure this information. Green algae and blue-green algae, like green plants, are photosynthetic deriving carbon and energy from the atmosphere and sunlight respectively. Minerals, deposited on the surfaces from dust, precipitation and condensation and nitrogenous deposits supply the algae with secondary nutrients and source of water. Algicides are therefore sometimes employed to keep the signs clear even though the material itself is not metabolised by the algae.

Claim: The surface will stay free from algal growth for more than 20 years.

In a tier 1 test only the basic principle can be proven. It has to be demonstrated in lab tests that there is growth of algae on an untreated surface whereas on a treated surfaces there is no growth.

Tier 2 testing, as mentioned before, would include also performance tests which include UV-irradiation, leaching, and temperature extremes that simulate in the laboratory the worst-case to be expected in the environment they are to be used.

Since the material is expected to perform in a very complex environment (nature as it is) it is useful when the claim can be substantiated with field test data. Also here a worst case scenario would be useful (wetness, light, abundance of algal growth in the surrounding test site) in an attempt to show a difference between treated and untreated surfaces.

Conclusion

What do biocides need to do in synthetic materials? They have to solve a problem. The first question therefore always is: what is the problem?

Can it be demonstrated that there is a problem?

Can it be demonstrated that there is less or no problem when a biocide is used?

Moving with the testing up from tier 1, to tier 2 and to tier 3 the more tailored to the field of application envisaged a test design has to be. While in tier 1 some existing standards are basically suitable when the biocide is tested in a suitable matrix with relevant organisms, tier 2 asks for more. Here, questions of long-term performance must be answered. What environmental conditions will the biocide-containing synthetic material be exposed to? More than one biocide can be used in a material, additives of all sorts can be mixed in to meet the requirements of long-term performance. At this stage testing has moved on from testing a basic principle (biocidal effect of a biocide) to testing a product in which its properties should be enhanced by the presence of a biocide.

Tier 3 investigates the benefit of the biocide in the product. After all, the protection of consumers and the environment must be weighed up against the gains in performance or economics that a biocide might bring.

Are there downsides of using the biocide? Are the benefits from the use of this biocide overbalanced by the downsides? These are difficult but important questions that deserve the full attention of any responsibly acting community.

Literature:

(1)   MIL-STD-810G, U.S. DEPARTMENT OF DEFENSE: Test method standard environmental considerations and laboratory tests, METHOD 508.6 ANNEX A, 31 October 2008; Downloaded from http://www.everyspec.com on 2010-04-14T9:38:54. MIL-STD-810G METHOD 508.6

(2)   MIL-STD-810G, U.S. DEPARTMENT OF DEFENSE: Test method standard environmental considerations and laboratory tests, METHOD 508.6 ANNEX B, 31 October 2008; Downloaded from http://www.everyspec.com on 2010-04-14T9:38:54. MIL-STD-810G METHOD 508.6

(3)   DIRECTIVE 98/8/EC OF THE EUROPEAN PARLIAMENT AND OF THE COUNCIL of 16 February 1998 concerning the placing of biocidal products on the market

(4)   ISO 22196 (2007): Plastics – Measurement of antibacterial activity on plastic surfaces

PROPOSED CHANGES TO THE REGULATION OF BIOCIDES AND THEIR POTENTIAL IMPACT

Geoff Wilson, Consultant
19 Elm Tree Road, Lymm, Cheshire, WA13 0ND, UK
Tel: (Mobile) 07947 666378 email: biocides@geoffandjill.wanadoo.co.uk

BIOGRAPHICAL NOTE

Geoffrey Wilson graduated as an analytical chemist from Leeds University in 1970. He has over 20 years experience working with biocides and non agricultural pesticides. He has mainly been employed by the UK Health and Safety Executive (HSE), part of which acts as the UK Competent Authority for Biocides (CA) and Geoff was a key member of the CA for a number of years. He has had two major secondments from HSE – one to the European Commission (1994-1996) where he worked on the text of the BPD and one to the OECD (2001-2003) where he ran the OECD's biocides programme. Since his retirement from the HSE in June 2008, he has worked as an independent consultant.

ABSTRACT

Geoffrey Wilson has over 20 years experience in the field of non-agricultural pesticides and biocides and was a key member of the UK Competent Authority for biocides for a number of years. Since his retirement, in June 2008, Geoff has operated as an independent consultant and the views expressed by him in this presentation are entirely his own and do not necessarily reflect those of any recognised organisation.

Geoff's presentation looks at the position with the current legislation on biocides (the Biocidal Products Directive, 98/08/EC (BPD)) and describes what stakeholders consider to be wrong with it. He then goes on to explain the new proposal from the European Commission and considers its key points which are:

- The use of a Regulation rather than a Directive
- The increase in scope
- The use of biocides
- Fees and charges
- Active substance exclusion criteria
- Comparative assessment
- Low risk biocidal products
- Data requirements
- Data waiving
- Data sharing
- Resolution of the 'Free Rider' issue
- The role of the European Chemicals Agency (ECHA)

He then explains what has happened in the European Council and the Parliament so far and ends the presentation with his thoughts as to whether or not this proposal for the Biocidal Products Regulation will be adopted as it stands as planned in January 2013.

Slide 1

PROPOSED CHANGES TO THE
REGULATION OF BIOCIDES & THEIR
POTENTIAL IMPACT

Geoffrey Wilson

Biocides in Synthetic Materials 2010
28-29 September 2010, Berlin

Slide 2

BIOCIDAL PRODUCTS DIRECTIVE (BPD)
98/8/EC

Came into force 14 May 2000

- Two main aims
  - High level of protection for humans & environment
  - Enable free trade within EC

- Biocidal Active Substances
  - Reviewed by rapporteur Member State on behalf of all EU. If a.s. is included in Annex I then it can be used in biocidal products for that product type

- Biocidal Products
  - Authorised in each Member State
  - Mutual recognition between Member States

Slide 3

> ## REVISION OF THE BPD
>
> ## · WHY?
>
> ## · Article 18.5 of BPD
>
> - requires the Commission to submit a report to the Parliament and Council on the implementation of the BPD together with any proposals for its revision (by May 2007)
>
> ## · It's not working very well!

Slide 4

> ## BPD - CURRENT SITUATION
>
> After approximately 10 years
>
> · Review programme very slow
>   - only about 40 a.s. on Annex I
>   - about 350 a.s. remaining
>
> · Very few products authorised
>
> · Transition period ends May 2010
>
> · High costs, particularly for SMEs

Slide 5

REVISION PROCESS

- Mini revision (COM/2008/618)
  - prolongs review prog. to 2014

- Major revision
  - MS consultation
  - Stakeholder consultation
  - Impact assessment study

- Proposal for a new Regulation
  - COM/2009/267
    (to come into force January 2013)

Slide 6

KEY ISSUES IDENTIFIED (1)

- Very high cost of compliance
  - data requirements excessive
- Data sharing and data protection
- Wide range of fees for a.s. review
- Scope not clearly defined
  - different interpretations in MS
  - foodstuffs included?
- Low risk substances and basic
  substances concepts not clear

Slide 7

KEY ISSUES IDENTIFIED (2)

- Discrimination between non-EU & EU manufacturers of treated articles
- Product authorisation process?
  - frame formulations concept
  - data requirements
  - fees for product authorisation
- Use phase of biocides
  - impact of other EU legislation
- **Simplification & Clarification**

Slide 8

BPD OR BPR?

Advantages of a Regulation
- Improves harmonisation
  - no differing implementations in MS
- Improves speed of implementation & reduces the administrative burden
- In line with COM Communication on simplification: REACH, PPP, etc.

Disadvantages of a Regulation
- More difficult to agree the text

Slide 9

MAIN CONTENT OF THE BPR

- Scope
  - Treated materials and articles
  - In situ generated actives
  - Food & food contact materials
  - Use of biocidal products
  - Definitions e.g. "Frame formulations"
- Fees
- Active substances
  - exclusion criteria
  - candidates for substitution & Comparative Asses.
- Biocidal products
  - Mutual recognition
  - Centralised procedure
  - Low risk products
  - Amendment of authorisations
- Data sharing & data protection
- ECHA involvement

Slide 10

SPECIFIC CONTENT OF THE BPR

Scope

- All treated articles and materials included in scope
- Imports of treated articles and materials may only contain approved products or active substances
- Labelling of treated articles and materials to be considered
- All 'in situ' active substances included in scope
- Biocidal products used in food contact materials included in scope
- Food and feed additives that are sufficiently covered by existing legislation are excluded from scope
- Biocidal products approved under the IMO to treat ballast water are considered to be authorised under the proposed Biocidal Products Regulation

Slide 11

SPECIFIC CONTENT OF THE BPR

Biocides use phase

- Pesticides Framework Directive
  - will not know outcome until after the adoption of BPR

- COM looking at the environmental impact of biocides at the use stage
  - workshop
  - technical study/questionnaire
  - MS input from their experience

Slide 12

SPECIFIC CONTENT OF THE BPR

Fees and charges

- Harmonised structure for fees and conditions of payment
- Reductions for SMEs
- Reduction for joint submissions
- Annual charge for all placing biocidal products on the market
- Fees reflect the required work

Slide 13

HARMONISED FEES STRUCTURE

Article 70 (2) (e)

"the structure and amount of the fees shall take account of the work required by this Regulation to be carried out by the Agency and the competent authorities and shall be fixed at such level as to ensure that the revenue derived from the fees when combined with other sources of the Agency's revenue pursuant to this Regulation is sufficient to cover the cost of the services delivered."

Slide 14

SPECIFIC CONTENT OF THE BPR

Active substances exclusion criteria

• Exclusion from Annex I entry for certain high hazard active substances (CMR, PBTs, endocrine disrupters etc.)

• These active substances may be included in Annex I only if at least one of the following conditions is met:

(a) the exposure of humans to that active substance in a biocidal product, under normal conditions of use, is negligible, in particular where the product is used in closed systems or strictly controlled conditions;

(b) it is shown that the active substance is necessary to control a serious danger to public health;

(c) it is shown that not including the active substance in Annex I would cause disproportionate negative impacts when compared with the risk to human health or the environment arising from the use of the substance and that there are no suitable alternative substances or technologies.

Slide 15

SPECIFIC CONTENT OF THE BPR

Active substances
- Comparative assessment in 2 stages
  - Active substance stage
    (largely hazard based)
  - Biocidal product stage
    (risk & socio-economic factors)

- Annexes IA & IB to be repealed

- Cumulative assessment

Slide 16

SPECIFIC CONTENT OF THE BPR

Centralised procedure - for low risk products or products based on new active substances
- Mutual recognition is automatic
- Proposed procedure:
  - Dossier submitted to ECHA
  - Validation by ECHA (60 days)
  - Dossier evaluation by CA (365 days)
  - Peer review through BPC (270 days)
  - ECHA opinion
  - Commission decision
  - All MS have to accept decision

Slide 17

SPECIFIC CONTENT OF THE BPR

Low risk biocidal product

- Safety margins for health and environment are 10 times greater than usually considered acceptable
- But not if CMR, PBT, endocrine disrupter, etc.
- Active substances in the biocidal product are contained in such a way that only a negligible exposure can take place under normal conditions of use and the product is handled under strictly controlled conditions during all other stages of its life cycle

Slide 18

SPECIFIC CONTENT OF THE BPR

Data Requirements

- Revised and a two tier system established
  - Tier II data may need to be submitted depending on the characteristics and intended use of the active substance, or on the conclusions of the Tier I data assessment

- Data waiving provisions strengthened
  - the data is not necessary owing to the exposure associated with the proposed uses
  - it is not scientifically necessary to supply the data
  - it is not technically possible to generate the data

Slide 19

SPECIFIC CONTENT OF THE BPR

Data sharing & data protection

- Mandatory data-sharing for test data on vertebrate animals
  - in line with REACH and PPP
  - compensation issues/use of arbitration

- Clarification of Article 12 of BPD
  - data protection to start at Annex I entry?

- Resolution of 'Free Rider' issue

Slide 20

SPECIFIC CONTENT OF THE BPR

Free Riders

- By 1 January 2015, manufacturers of active substances to submit a dossier or a letter of access to ECHA
- ECHA to publish a list of manufacturers
- Mandatory data and fair costs sharing to apply to the whole dossier
- Biocidal products containing existing active substances of manufacturers not on the ECHA list shall not be placed on the market after 1 January 2015
- Disposal, storage and use of existing stocks allowed until 1 January 2016
- Competent Authorities to enforce

Slide 21

SPECIFIC CONTENT OF THE BPR

ECHA involvement

- **Coordination role for evaluation of active substances**

- **Community authorisation opinion**

- **Dispute settlement**
  - **mutual recognition**
  - **data requirements**

- **Scientific and technical support**

Slide 22

EUROPEAN COUNCIL ACTIVITY

Swedish Presidency (July to Dec. 2009)

- **Thematic approach**
  - **Centralised authorisations**
  - **Low risk biocidal products**
  - **Exclusion criteria for active substances**
  - **Treated articles**
  - **Comparative assessment**
  - **Harmonised fee structures**

- **Policy debate in December Council**

Slide 23

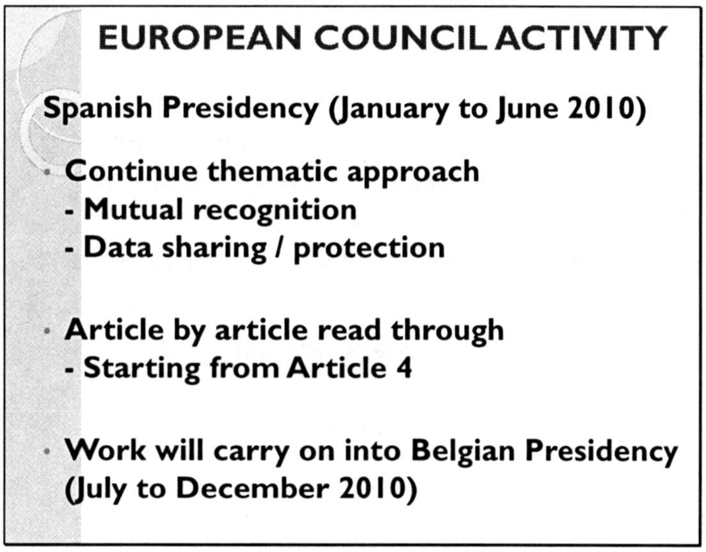

EUROPEAN COUNCIL ACTIVITY

Spanish Presidency (January to June 2010)

- **Continue thematic approach**
  - Mutual recognition
  - Data sharing / protection

- **Article by article read through**
  - Starting from Article 4

- **Work will carry on into Belgian Presidency (July to December 2010)**

Slide 24

EUROPEAN PARLIAMENT ACTIVITY

- **Rapporteurs appointed**
  - ENVI, IMCO, ITRE
- **Rapporteurs prepared reports**

- **Many amendments tabled**

- **Votes in each Committee**

- **Vote in plenary**

Slide 25

NEXT STEPS IN PROCESS

- Negotiations in Council of Ministers and European Parliament will continue
  - Co-decision process
  - Many amendments to be considered

- Second reading agreement in 2011?

- Final adoption in January 2013?

Slide 26

GEOFF THOUGHTS (1)

- It's very ambitious:
  "The new regulation increases the protection of health and the environment, while being more efficient at the same time, notably through the active involvement of ECHA. It will retain the two-step authorisation process brought in by the current directive, whereby active substances are first tested and approved in a community list (known as the Annex I) with subsequent authorisation of a product containing the active substance"

Slide 27

GEOFF THOUGHTS (2)

Simplified and clearer than BPD?

- **Scope is increased – all treated articles**
- **Text covers 193 pages**
  (BPD is 63 pages)
- **Text needs to be positively agreed by 27 MS**
- **Member States to accept less data**
- **Role played by ECHA**
- **Adoption anticipated in January 2013**

Slide 28

CONCLUSION

Slide 29

FURTHER INFORMATION

European Commission website:

http://ec.europa.eu/environment/biocides/revision.htm

Questions to: ENV-Biocides@ec.europa.eu

SURFACE BIOCIDES FOR FOOD CONTACT APPLICATIONS

Rachida Semail, Partner
Keller & Heckman LLP
Avenue Louise 523, B-1050 Brussels, Belgium
Tel: 0032 (2) 645 50 94   Fax: 0032 485 70 84 39   email: semail@khlaw.be

BIOGRAPHICAL NOTE

Rachida Semail, French qualified attorney, is a Partner in the Brussels office of Keller and Heckman LLP, where she heads the Food Packaging Practice. Ms. Semail advises clients in a broad range of issues, with particular emphasis on establishing a suitable status for articles and materials intended for food contact at EU and national level, assistance in case of enforcement issues, product recalls and liability. Ms. Semail has extensive experience on subject matters not harmonized at the EU level, as is the case in much of the food contact sector, where mutual recognition is of paramount importance.

Ms. Semail also has extensive expertise on matters of food law, including in the area of enzymes, flavourings, and food additive clearances, of particular relevance to active packaging. She also regularly assists clients on clearance and compliance issues relating to feed, cosmetics, drugs and medical devices.

Prior to joining Keller and Heckman, Ms Semail worked for a French law firm specializing in EU and French food law. She was an intern with the legal service of the European Commission in Brussels.

ABSTRACT

The presentation will provide a general overview on how biocides used in food contact are regulated in the EU and will focus on surface biocides.

Slide 1

Slide 2

Agenda

- Different possible uses of biocides in food contact materials
- Current applicable requirements for food contact surface biocides
- Future changes for food contact surface biocides

2 | www.khlaw.com | KELLER AND HECKMAN LLP   Copyright © 2010

Slide 3

Agenda

- Different possible uses of biocides in food contact materials
- Current applicable requirements for food contact surface biocides
- Future changes for food contact surface biocides

3 | www.khlaw.com | KELLER AND HECKMAN LLP   Copyright © 2010

Slide 4

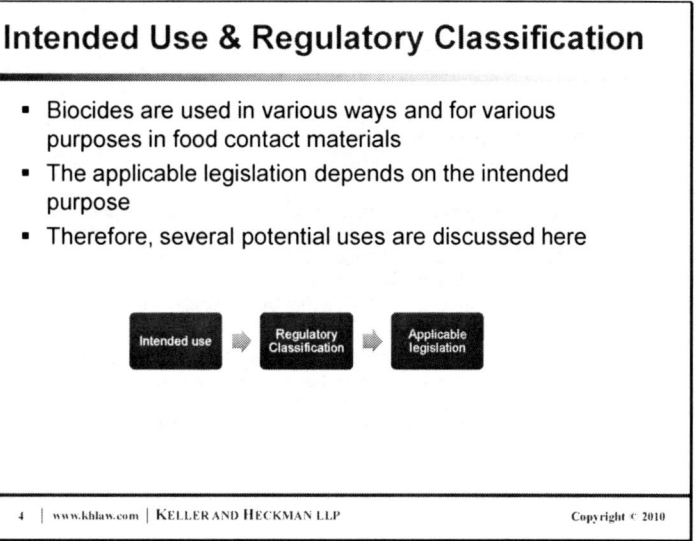

Slide 5

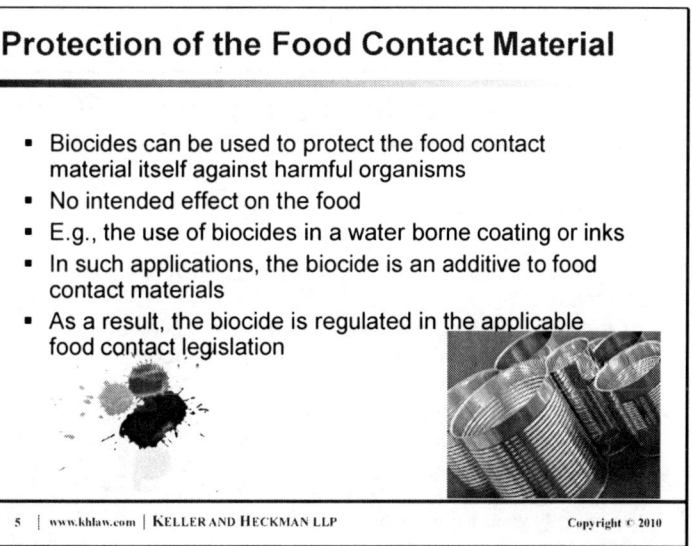

Slide 6

Protection of the Packed Food

- Biocides can be added to packaging with the intention to have them migrate into food
- The biocide will protect the food itself against harmful organisms
- E.g., Addition of benzoic acid to a multilayer, which will migrate into the food when the packaging is in contact with the food. Benzoic acid is intended to improve the conservation of the food (preservative)
- This use is regulated by the Active and Intelligent Packaging Regulation 405/2009
- The biocide will be in this case regarded as a food additive
- The biocide must be listed in the EU food additives positive list currently set out by Directive 95/2 (that has been replaced by Regulation 1333/2008 although the positive lists of the old Directive are for the time being still applicable)

6 | www.khlaw.com | KELLER AND HECKMAN LLP   Copyright © 2010

Slide 7

Biocides in Cleaning Liquids for FCM

- Biocides can be used to protect the surface of the food packaging itself against harmful organisms
- Not intended to migrate into food
- E.g., Use of biocides in cleaning solvents for kitchen surfaces
- As the biocide is not used/added in the production of the food contact material, it is not covered by the food contact legislation
- Status of such product is that of a biocide as defined in the Biocidal Products Directive 98/8
- The use of the biocide must meet the requirements of the Biocidal Products Directive (Product-type 4: Food and feed area disinfectants)

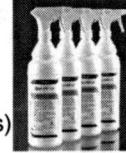

7 | www.khlaw.com | KELLER AND HECKMAN LLP   Copyright © 2010

Slide 8

Protection of the Surface of a FCM

- Biocides can be added to the packaging to protect the surface of the food contact material against harmful organisms, which might grow for example in scratches of the food contact material
- Not intended to have an effect in the food
- E.g., Use of silver molecules in conveyor belts
- Current status of such biocide is that of an additive to food contact materials
- Its status in the food contact legislation is complicated (will be discussed in next part of presentation)

8 | www.khlaw.com | KELLER AND HECKMAN LLP   Copyright © 2010

Slide 9

Surface Biocides in Fridges

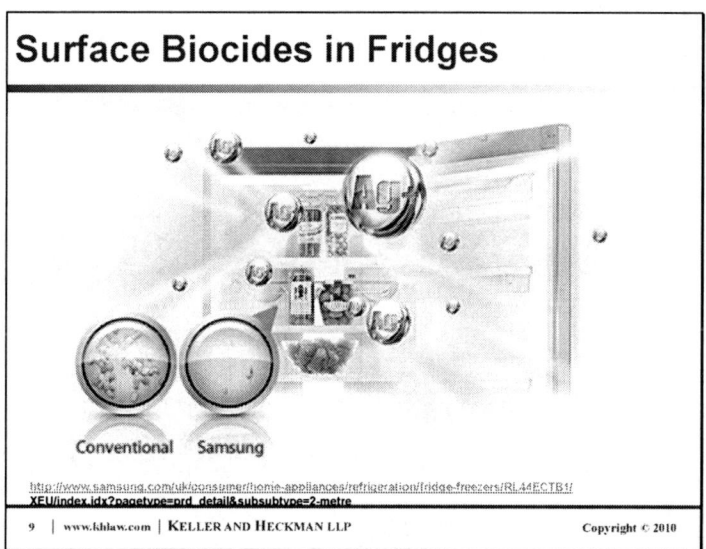

Conventional   Samsung

http://www.samsung.com/uk/consumer/home-appliances/refrigeration/fridge-freezers/RL44ECTB1/XEU/index.idx?pagetype=prd_detail&subsubtype=2-metre

9 | www.khlaw.com | KELLER AND HECKMAN LLP   Copyright © 2010

Slide 10

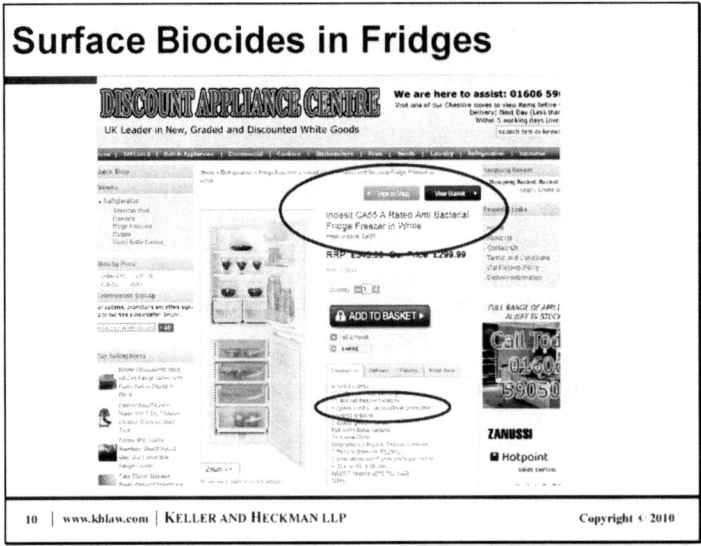

Slide 11

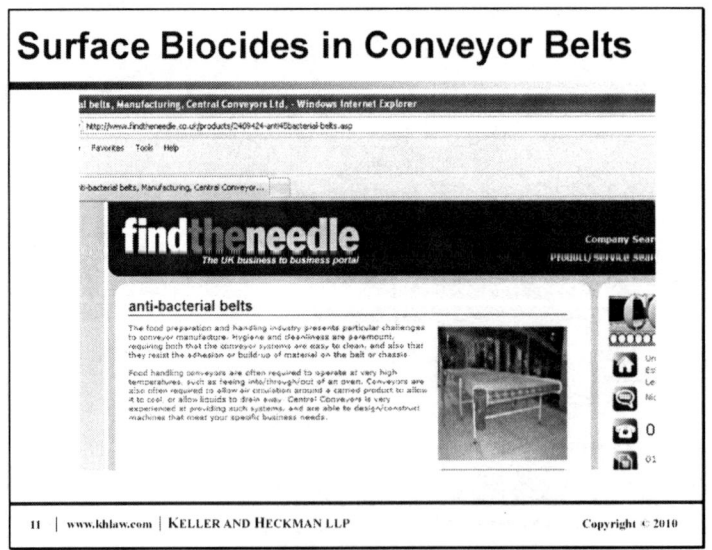

Slide 12

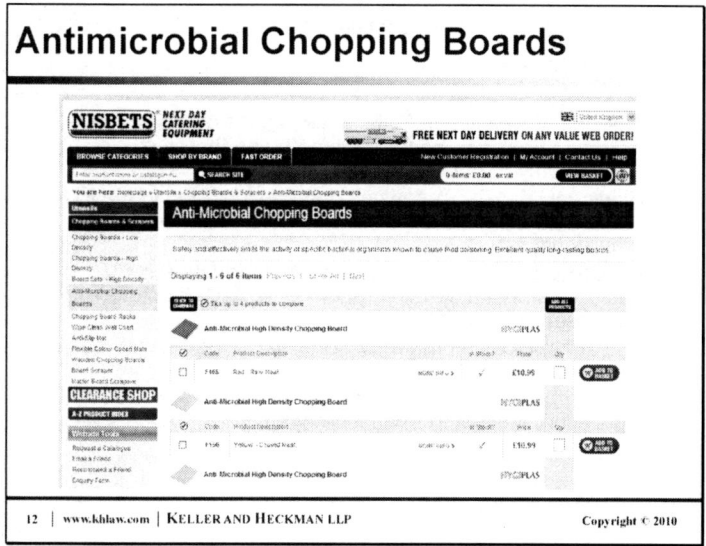

Slide 13

Agenda

- Different possible uses of biocides in food contact materials
- **Current applicable requirements for food contact surface biocides**
- Future changes for food contact surface biocides

13 | www.khlaw.com | KELLER AND HECKMAN LLP   Copyright © 2010

Slide 14

Surface Biocides Under The Food Contact Legislation

- The use of biocides in FCM is excluded from the Biocidal Products Directive, so for the time being they are regulated by the FC legislation
- Since Jan. 1, 2010, additives must be included in the EU positive list of additives when used in plastics or in the provisional list
- No surface biocides are currently included on the EU positive list; but a certain number of surface biocides are on the provisional list
- The substances on the provisional list all have a favorable opinion from the EFSA
- Substances on the provisional list are subject to national legislation, subject to mutual recognition

14 | www.khlaw.com | KELLER AND HECKMAN LLP   Copyright © 2010

Slide 15

Surface Biocides Under The Food Contact Legislation

Surface biocides on the Provisional list:
- 20%w/w Silver chloride coated onto 80% (w/w) titanium dioxide
- Silver-containing glass (silver-magnesium-calciumphosphate-borate)
- Silver containing glass (silver-magnesiumaluminium-phosphate-silicate), silver content less than 2%
- Silver containing glass (silver-magnesiumaluminium-sodium-phosphate-silicate-borate), silver content less than 0.5%
- Silver containing glass (silver-magnesium-sodiumphosphate), silver content less than 3 %
- Silver sodium hydrogen zirconium phosphate
- Silver Zeolite A (Silver zinc sodium ammonium alumino silicate), silver content 2-5%
- Silver-zinc- aluminium – boron – phosphate glass, mixed with 5-20% barium sulphate, silver content 0,35 – 0,6 %
- Silver zinc zeolite A (silver zinc sodium alumino silicate calcium metaphosphate), silver content 1 -1,6 %
- Silver zinc zeolite A (silver-zinc sodium magnesium alumino silicate calcium phosphate), silver content 0,34-0,54 %

8th update (v2): 27/05/2010

15 | www.khlaw.com | KELLER AND HECKMAN LLP   Copyright © 2010

Slide 16

Surface Biocides Under The Food Contact Legislation

- Political debate surrounding their inclusion in the EU positive list
  - Issue arose because of Triclosan
  - Triclosan is an organic, low Mw surface biocide
  - Although it is not intended to migrate to the food, in practice, it is likely to migrate into the food when in contact
  - Strong objections from national competent authorities because of Triclosan while objection does not seem to be justified for the silver-based biocides

16 | www.khlaw.com | KELLER AND HECKMAN LLP   Copyright © 2010

Slide 17

Surface Biocides under the Food Contact Legislation

- The surface biocides were included in some draft amendments of the Food Contact Plastics
- There were attempts to reach a consensus at EU level by restricting their use to :
  - Industrial environments (non-consumer products)
  - For repeated use applications only
- However, they were always removed before the draft was finalized (to avoid that no favorable voting was obtained for the draft legislation)

17 | www.khlaw.com | KELLER AND HECKMAN LLP   Copyright © 2010

Slide 18

Triclosan

- The petitioner of triclosan has withdrawn the substance (although a positive opinion of the EFSA was available)
- As a result, triclosan was removed from the provisional list following the publication of the Commission Decision 2010/169 of 19 March 2010
- A transitional period is foreseen by the Decision: plastic materials and articles manufactured with triclosan and placed on the market before 1 November 2010, may continue to be marketed until 1 November 2011, subject to national legislation.

Despite of the withdrawal of triclosan, the other surface biocides are not (yet?) included in the food contact legislation!!!!

18 | www.khlaw.com | KELLER AND HECKMAN LLP   Copyright © 2010

Slide 19

Today's Use of Surface Biocides

- Surface biocides can be found in many applications
- The use of these applications is widely marketed
- Some Member States have a strong opposition regarding the use of surface biocides; Using surface biocides may raise no issues in other Member States
- However, enforcement seems to be so far limited

19 | www.khlaw.com | KELLER AND HECKMAN LLP   Copyright © 2010

Slide 20

Agenda

- Different possible uses of biocides in food contact materials
- Current applicable requirements for food contact surface biocides
- **Future changes for food contact surface biocides**

20 | www.khlaw.com | KELLER AND HECKMAN LLP   Copyright © 2010

Slide 21

New Biocides Regulation

- A new Biocides Regulation is expected to enter into force by 1 January 2013
- By contrast to the current Biocidal Products Directive, the use of surface biocides in food contact materials are expressly covered by this new biocides Regulation
- Does this mean that the biocides are no longer regulated by the food contact legislation but by the Biocide Regulation only. Or by both regulations?

21 | www.khlaw.com | KELLER AND HECKMAN LLP   Copyright © 2010

Slide 22

So What then?

Which substances can be used in the future?
- The substances mentioned in the Biocidal Products Regulation?
- The substances mentioned in the Food contact Regulation?
- Substances mentioned in both sets of legislation?
- Substances mentioned in either Biocidal products Regulation or the Food Contact Legislation?

22 | www.khlaw.com | KELLER AND HECKMAN LLP   Copyright © 2010

Slide 23

Evaluation of Dossiers

- Petitions of the Food Contact plastics are reviewed by the EFSA
- Petitions of non-harmonized Food Contact substances are reviewed by the Member States
- European Chemicals Agency (ECHA) will take over the coordination of the EU evaluation of biocides and will play key role in the centralised authorisation of low-risk and of new substances.
- Question: Must dossiers for biocides that already have a favorable EFSA opinion be re-submitted in order to be included in the Biocidal Products Regulation?

23 | www.khlaw.com | KELLER AND HECKMAN LLP   Copyright © 2010

Slide 24

Conclusions

- It is important to distinguish the different uses of biocides in combination with food contact materials, as different sets of legislation apply
- Under the current legislation, surface biocides are regulated by the food contact legislation
- For plastics, 10 substances are listed in the provisional list. Some resistance in some Member States exists for the use in food contact materials
- In the future (after 2013) surface biocides will be included within the scope of the Biocidal Products Regulation (food contact materials no longer exempted) and the legal framework will need to be clarified for surface biocides for use in food contact applications

24 | www.khlaw.com | KELLER AND HECKMAN LLP   Copyright © 2010

Slide 25

Thank you

Rachida Semail
Partner
Keller and Heckman LLP
Avenue Louise 523
1050 Brussels, Belgium
32(2) 645-5094
semail@khlaw.be

www.khlaw.com

Washington, D.C. • Brussels • San Francisco • Shanghai

KH KELLER AND HECKMAN LLP
Serving Business through Law and Science®

DIETARY RISK ASSESSMENT OF BIOCIDES - ASSESSING EXPOSURE OF LIVESTOCK ANIMALS TO BIOCIDES

Isabel Günther

Bundesinstitut für Risikobewertung (Federal Institute for Risk Assessment)

Thielallee 88-92, 14195 Berlin, Germany

Tel: 0049 (30) 18412-3956 email: Isabel.Guenther@bfr.bund.de

BIOGRAPHICAL NOTE

Isabel Guenther of the Federal Institute for Risk Assessment (BfR) in Berlin, Germany, began working for her institute in July 2008. Her initial work focused on the assessment of pesticide and biocide residues in food. Current responsibilities include the preparation of overall assessment reports as part of the EU review programme for biocidal active substances. She is further involved in commenting the draft biocide regulation.

In May of 2009 she took the position of chair of the EU working group DRAWG (Dietary Risk Assessment Working Group), which is charged with developing harmonised guidance for the assessment of livestock and food exposure to biocides. Ms. Guenther is also a member of the CVMP-BTM Working Group (Committee for Medicinal Products for Veterinary Use / Biocides Technical Meeting) which is developing guidance for MRL-setting for biocidal active substances.

Ms. Guenther holds a Bachelor of Science in Food Science from the University of Georgia (USA) and a Master of Science in Nutrition from the University of Massachusetts (USA) and can be reached at Isabel.Guenther@bfr.bund.de.

ABSTRACT

A guidance document has been developed for the external exposure assessment of livestock animals exposed to biocides for the purpose of evaluating residues in food produced by these animals. The guidance document is part of a set of papers providing guidance on the dietary risk assessment of biocides. Five groups of intended uses were identified by way of which livestock animals can be exposed to biocidal active substances: treatment of animal housing, treatment of feedstuff and drinking water or of storage facilities, treatment of materials that livestock animals may come in contact with, direct treatment of livestock animals, treatment of aquaculture. For each of these groups, example exposure estimations are provided that take into account the oral, dermal and inhalation exposure routes. The assessment approach follows a step-wise concept in which the outcome of each estimation is compared to a set threshold value that triggers the next step. Only external exposure assessment is described, not taking into account factors such as absorption or metabolism. These factors, which are part of internal exposure assessment, are considered in a separate guidance document.

+++ PAPER UNAVAILABLE AT TIME OF PRINT +++

FORMALDEHYDE RELEASERS AS AN EXAMPLE OF IN SITU GENERATION OF ACTIVE SUBSTANCES

Sylwester Huszał, Robert Chrobak
Office for Registration of Medicinal Products, Medical Devices and Biocidal Products
Zabkowska 41, PL-03-736 Warsaw, Poland
Tel: +48 22 4921 635 Fax: +48 22 4921 649 email: **sylwester.huszal@urpl.gov.pl**

BIOGRAPHICAL NOTE

Sylwester Huszal received a Ph. D. in Chemistry from Warsaw University in 2006. In 2007 he started to work in the Office for Registration of Medicinal Products, Medical Devices and Biocidal Products as a Head of Section for Assessment of Physical-Chemical Documentation of Biocidal Products. Within few months he became a Deputy Head of the Biocidal Products Assessment Unit and in 2009 he became a Head of the Unit. He regularly represents Polish Competent Authority at Technical Meetings on Biocides and is responsible for the coordination of practical aspects of assessment of active substances and biocidal products in Poland.

Robert Chrobak received a Ph. D. in Chemistry from Warsaw University in 2007. In the same year he started to work in the Office for Registration of Medicinal Products, Medical Devices and Biocidal Products in the Section for Assessment of Physical-Chemical Documentation of Biocidal Products. In 2009 he became a Head of this Section. He is responsible for the practical aspects of assessment of active substances and biocidal products in Poland, especially in the field of identity, physical-chemical properties and methods of analysis. As a Project Manager he is coordinating the evaluation process of 5 active substances allocated to Poland.

ABSTRACT

Formaldehyde-releasing compounds, also often called a formaldehyde donors are highly effective biocides used primarily to prevent unwanted microbial spoilage of water based products. They have proved to be very important for the protection of a wide range of materials. Important areas of formaldehyde releasers use include the preservation of aqueous products such as paints, sealants, emulsions, industrial water treatment, adhesives, plastics, paper chemicals, aqueous mineral slurries and household goods. The hydrolysis of formaldehyde releaser in water is an equilibrium reaction and the dilution factor is the main determinant for the amount of formaldehyde released from formaldehyde releaser. The paper describes briefly current status of formaldehyde releasers evaluation as active substances to be used in biocidal products as well as documents prepared for the purposes of evaluation process of in situ generated active substances.

INTRODUCTION

First attempt to the subject of in situ generated active substances was document "**Guidance document agreed between the Commission services and the competent authorities of Member States regarding the in-situ generation of active substances and related notifications**"- discussed in September 2001 by the European Commission and the Member States - and officially published in 2002. This very short document is still very helpful for Applicants and Member States during the discussions how to deal with notifications of active substances and/or authorization of biocidal products which may be called/may contain releasers of in situ generated active substances. For the first time a substances/mixtures which are placed on the market and may release active substance were called 'precursor' chemicals. In mentioned paper you may find information about already recognized examples of in situ generation of active substances, their classification as active substances, requirements for data for those substances and specific cases which need more attention during dealing with together with proposals how to resolve problems that have appeared during the discussions concerning in situ generation of active substances in biocidal products. In Annex I to that document you may find CEFIC Information on Formaldehyde Releasing Biocides in the context of the EU Biocidal Products Directive (98/8/E). This very important group of chemicals that are commonly used as active substances in biocidal products build up a group of more than 18 notified biocides which are used in a variety of different product types. This results in industry proposal how to deal with such a wide group of biocides.

Some very important information on in situ generation and on site formulation of biocides may be find in **MANUAL OF DECISIONS FOR IMPLEMENTATION OF DIRECTIVE 98/8/EC CONCERNING THE PLACING ON THE MARKET OF BIOCIDAL PRODUCTS** which was last modified In July 2008. In Chapter 2.2 nine real examples of in situ generation of active substances is detailed examined together with final conclusion agreed between all Member States. Also description of on site formulation and use of biocidal products is presented in that document that is in use during the review program and is extremely helpful for all parties during the authorization of biocidal products in all Member States.

In a couple of years common knowledge of Member States gained during the review program resulted in UK proposal for data requirements for the in situ generation of active substances for biocides prepared by the UK Competent Authority in early 2007. This proposal extends official document from 2002 and its main aims were:

- to describe the principles of in situ generation of biocidal active substances;
- to describe as many examples of in situ generation as it is possible;
- to propose a list of data requirements for different in situ generated active substances;
- to propose how o evaluate those substances and how to listed them in Annex I of the directive 98/8/EC.

In that document you may find new, slightly changed definitions of an in situ generated active substance and precursor, description of recognized methods for the generaton of in situ biocidal active substances like: chemical reaction, electrolysis, UV radiation to generate free radicals, release of biocidal active substances under specific use conditions, generation of unstable or highly reactive active substances which cannot be isolated. Described in document new approach to data requirements for dossier evaluation for Annex I inclusion in detailed way illustrates approach of Member States to evaluation of dossiers for in situ generated active substances and their precursors e.g. what can be waived in documentation and in what way it should be argument; what to do when required data cannot be generated due to the properties of active substance, what kind of data may be required for hazard assessment.

As previously Annex I of that document is dedicated to formaldehyde releasers due to the fact that CEFIC would like to see a specific guidance on formaldehyde releasers as they form a large category of in situ generated active substances. This document was commented by various Member States but due to fact that UK Competent Authority declared that have no resources for further work on that document it was decided that CEFIC will lead further work on document on in situ generated active substances.

As a consequence of that fact in September 2008 new industrial (CEFIC) proposal of the guidance document regarding active substances generated in situ and their evaluation under Directive 98/8/EC was sent by the Commission to the Member States for consultation. Several countries have send their comments and new version of the document was prepared in July 2009. It has not been finished yet and is still under consultation with the Member States and the Industry.

The draft guidance is the proposal resulting from the experiences from review program and is addressing the most of the problems that Member States or the Industry had to face during this process. Proposed document does not address technical issues of in situ generaton of the active substances, probably because of complexity of the issues related to in situ generation.

This paper contains information on principles of in situ generation of active substances. It should be stressed that some products containing substances (precursors) with no biocidal activity which during the normal use release an active substance do not fit exactly to definitions of active substance and biocidal product from the Directive 98/8/EC (biocidal product should contain one or more active substances when it is placed on the market). However, the "common sense" approach and the following decisions based on the purpose for placing of this kind of products on the market causes them to fall within the scope of Biocidal Products Directive. The draft proposal is also defining the terms related to in situ generation of actives. Due to mentioned upper complexity of this issue, appropriate definitions, such as in situ generated active substance, precursor active substance (taking into the account different kinds of precursors – an active, non-active ones, *etc.*) are clarifying the situation and are helping with the understanding of the general approach to in situ generated active substances. Additionally, the list of identified and used methods for the generation of active substances is presented but since it is not closed list, further methods can be also added to the document. The guidance is also reflecting on the data requirements for the evaluation. The information presented is rather a collection of suggestions from Industry point of view than the data requirements. As a consequence establishing of the core data set covering requirements for all methods of generation and different types of precursor active substances would be extremely difficult, if possible at all. The data set appropriate for evaluation and risk assessment for precursor and eventually (separately) for in situ generated active substance (like formaldehyde core dossier provided as an addendum to all formaldehyde releasers dossiers) must be addressed on case by case basis. Finally, the draft guidance document contains the proposal of Annex I entries which can be different than the entries for common active substances, depending on the kind of precursor and method of in situ generation of active substance and is correlated with appropriate precursor type.

The structure of a guidance document is opened, the presented information is supported by the real examples what makes it more friendly for the reader and additionally it can be developed and extended without changing of its general structure. Parties interested in participation in developing of that document are welcomed to contact Commission or CEFIC directly.

Identification of formaldehyde-releasing substances

A large number of formaldehyde releasers is widely used in the industry as bactericides, but they can differ from each other in:

- the chemical identity (i.e. O-formal and N-formal compounds);
- the composition (pure single compound, reaction mixture with two or more main components in equilibrium, reaction mixture with main component(s) and several by-products (oligomers));
- the kinetics of formaldehyde release (fast and slow release);
- the formaldehyde release (hydrolysis to educts or hydrolysis via stable intermediates);
- amount of releasable formaldehyde.

It was obvious that a decision at EU-level - harmonized as far as scientifically possible - regarding the data package (of course reduced if it would be possible) necessary for a group of similar substances - to be seen as precursors for formaldehyde, based on a structured procedure and long before the time of submitting the dossiers - would be an advantage for both parties involved, for the responsible authorities as well as for the applicants. That need for that harmonization during the evaluation of formaldehyde releasers was a starting point for further discussions and creation of FREG.

Formaldehyde Releasers Evaluation Group – a way forward in HCHO releasers evaluation

In autumn 2007 based on the proposal of Austria and Poland Formaldehyde Releasers Evaluation Group (FREG) which contains the Rapporteur Member States to which the formaldehyde releasing substance dossiers have been submitted was established. First meeting of FREG members titled 'Evaluation of Dossiers of Formaldehyde Releasers' took place on 11-12 October 2007 in Warsaw. In meeting participated representatives from Austria, Germany, Poland and Spain. Invitation to the meeting was also send to UK Competent Authority, but due to the other circumstances experts from UK could not come to Warsaw. To allow them to take part in discussion a teleconference with UK experts was organized during the first day of the meeting. During the meeting it was agreed that the establishment of a Formaldehyde Releasers Evaluation Group (FREG) is a necessary step for a better and more fruitful cooperation between Competent Authorities of Member States responsible for the evaluation of dossiers for formaldehyde releasers. All Member states present at the meeting, (including the UK), expressed willingness in joining the FREG.

During the first meeting participants discussed a way forward for harmonized evaluation of the formaldehyde releasing substances' dossiers, proposals for data requirements for the in situ generation of active substances for biocides. Also Annex I listing of formaldehyde releasers was deeply discussed during the meeting. Two proposals were presented. It was very hard to decide between 'Formaldehyde as released from [active substance]' and '[Active Substance] releasing formaldehyde'. The second proposition was more in favour and additionally it was agreed also to propose listing of '[Active Substance]' in the case where efficacy is due to the parent substance (not only due to the presence of formaldehyde released from active substance).

Because of high workload in all members of FREG it was finally decided not to allocate the evaluation of all formaldehyde releasers dossiers to one Member State and based on that German CA will decide to join the FREG and contributing to evaluation of formaldehyde releasers after receiving of formaldehyde dossier. German participation in FREG was acknowledged by other members of group as very important and helpful. Accordingly for better cooperation and coordination of future work several agreements were set as a result of the first meeting of FREG:

- formaldehyde releasers (FAR) dossiers format agreed by the present participants;
- Completeness Check approach and further consultations on that step, especially interchange of opinions on the completeness of the formaldehyde dossier accompanying the dossiers for all releasers;
- exchange of data summarizing the availability of data for formaldehyde releasers together with references to existing studies for formaldehyde;
- approach on formaldehyde dossiers in case where there will be no submission of dossier for formaldehyde to Germany.

Second workshop on 'Evaluation of Dossiers of Formaldehyde Releasers' took place on 14-15 January 2008 also in Warsaw. In expert meeting participated representatives from Austria, Germany, Hungary, Poland, Spain and United Kingdom.

The main topics discussed during that meeting were:

- submission of formaldehyde dossier to Germany together with dossier for a new active substance, a formaldehyde releaser. It was agreed that as soon as it will be possible Germany will perform the Completeness Check and the evaluation of the formaldehyde core dossier. This work will be done in close cooperation with other members of FREG.
- agreements for the formaldehyde core dossier and formaldehyde releasers. All participants agreed that dossiers for formaldehyde releasers are "quasi' complete and taking of further steps of evaluation will depend on each member of FREG.
- active substances as a UVCB substances mixtures. Based on the available information and experience it was agreed that UVCB substances should be described by the manufacturing method and that studies should be performed for the active substance as manufactured.

Until now (August 2010) next 4 meetings of FREG members took place. Two of them were organized by Federal Institute for Risk Assessment in Berlin (3rd meeting- 19-20.06.2008 and 5th meeting- 14-15.10.2009) and two of them by the Office for Registration of Medicinal Products, Medical Devices and Biocidal Products in Warsaw (4th meeting- 22-23.01.2009 and 6th meeting- 10-11.05.2010)

During those 3 years of cooperation the full evaluation of the formaldehyde core dossier was performed by Germany and circulated within the FREG group states for comments which were received and discussed during experts meetings in Berlin and Warsaw. Due to the some deficiencies in the environmental part of formaldehyde dossier applicants were asked to send new data to fulfil requirements of FREG members. The agreed letter containing the demand was sent to all applicants of formaldehyde core dossiers in order to achieve a harmonized approach for all the formaldehyde releasers dossiers and to avoid that participants and the applicant will be treated in a different manner. It was agreed that the problem of labelling of products containing formaldehyde releasers should be addressed in the Product Authorisation & Mutual Recognition Facilitation Group on CA-level.

The rate of HCHO release strongly depends on the speed of hydrolysis. To provide a fast effect a fast release of formaldehyde is required while for a long protection a slow release is needed. The scientific justification to define a compound a slow or fast releaser depends on stability of that formaldehyde releaser. After detailed discussion on many difficulties in distinguishing between slow and fast releasers it was underlined that the rate of formaldehyde emission depends on pH, releaser concentration and temperature. Also, experts on toxicology and ecotoxicology have different understanding of fast and slow releasing. It was concluded that it is difficult to unequivocally define fast/slow formaldehyde releasers and exposure should be considered step-by-step approach for each releaser.

During the 5th Meeting of FREG members group was joined by Italy – a reporter member state for TETRAKIS active substance (on the behalf of Malta) which might be regarded as a formaldehyde releaser. It was agreed that TETRAKIS formally do not belong to the formaldehyde releasers because of its mode of action, but on the other hand formaldehyde is contained in the active substance and additionally is a metabolite of it. Therefore, data for formaldehyde are necessary for the evaluation of TETRAKIS.

It was also decided that the core dossier will be updated every 2nd year with an addendum containing text and references as long as no substantial changes will be made.

Two document prepared by Polish experts for FREG meetings: Document for discussion of emission scenarios for biocides used as in-can preservatives (PT6) and Document for risk assessment of formaldehyde releasers used for the preservation of metalworking fluids (PT13) were distributed before 6th meeting of FREG and after detailed discussion were sent to the Technical Meeting for further consideration. Based on cooperation of FREG members Poland prepared CA Report for the first formaldehyde releaser (DMDMH) that should be submitted in a very near future (hopefully in September 2010) to the other Member States for consultation period.

Polish proposal for the next FREG Meeting is May 2011 in Warsaw. In the case of some very important problems that may arise suddenly the next FREG Meeting may be arranged earlier and hosted by Germany or Spain.

Table 1. List of formaldehyde releasers under evaluation

Member States responsible for evaluation	Formaldehyde releaser	CAS number
Poland	(Ethylenedioxy)dimethanol **EG-formal**	to be clarified
	Methenamine 3-chloroallylochloride **CTAC**	4080-31-3
	2,2',2"-(Hexahydro-1,3,5-triazine-1,3,5-triyl)triethanol **HHT**	4719-04-4
	1,3-Bis(hydroxymethyl)-5,5-dimethylimidazolidine-2,4-dione **DMDMH**	6440-58-0
	7a-Ethyldihydro-1H,3H,5H-oxazolo[3,4-c]oxazole **EDHO**	7747-35-5
	cis-1-(Chloroallyl)-3,5,7-triaza-1-azoniaadamantane chloride **Cis-CTAC**	51229-78-8
Austria	a, a ', a "-Trimethyl-1,3,5-triazine-1,3,5(2H,4H,6H)-triethanol **HPT**	to be clarified
	3,3'-methylenebis[5-methyloxazolidine] **MBO**	to be clarified
	N,N-Methylenebismorpholine **MBM**	5625-90-1
	Sodium hydroxymethyl glycinate, **Glycinate**	70161-44-3
Spain	Tetrahydro-1,3,4,6-tetrakis-(hydroxymethyl)-imidazo[4,5-d]imidazole-2,5(1H,3H)-dione **TMAD**	5395-50-6
	2-Bromo-2-nitro-1,3-propanediol **BRONOPOL**	52-51-7
UK	4,4-Dimethyloxazolidine **DMO**	51200-87-4
	(Benzyloxy)methanol **BHF**	14548-60-8
Germany	Reaction products of ethylene glycol, urea and paraformaldehyde **EUF**	to be clarified
Italy	Tetrakis(hydroxymethyl)phosphonium sulphate(2:1) **TETRAKIS**	55566-30-8

BIOAEROSOLS AS PART OF THE "SICK BUILDING SYNDROME" AND "BUILDING RELATED ILLNESS": 2 CASE STUDIES

Carlos A Rocha*, Ricardo J Silva*, Aurora E Monzón**, Juvenal Alfonzo**, Nayila A Baez, Evelys V Villarroel, María G Quintero***

*Universidad Simón Bolívar

Valle de Sartenejas, Caracas, Venezuela, Tel: 58 (416) 9058135 Fax: 58 (212) 9063064

email: crocha@usb.ve

** Electricidad de Caracas, San Bernardino, Caracas, Venezuela,

*** Hospital Universitario de Caracas, Caracas, Venezuela.

BIOGRAPHICAL NOTE

Prof. Carlos Rocha was born in Caracas, Venezuela, in 1959. He obtained his Bachelor degree with Honor in Biomedicine at Northeastern University, Boston, USA in 1982 and also in Clinical Psychology at Universidad Central de Venezuela, Caracas, Venezuela, in 1992. He got his *Magister Scientiarum* degree in Microbiology at The Venezuelan Institute of Scientific Research (IVIC), Caracas, Venezuela, in 1987 and his *Ph.D.* in Microbiology at Simón Bolívar University, Caracas, Venezuela, in 1993. Currently, he is an Associate Professor of Microbiology in the Cell Biology Department at Universidad Simón Bolivar and an invited Professor of Microbiology at the Faculty of Medicine at the Universidad Central de Venezuela. He has published scientific research papers mainly related to the subjects of Air Microbiology and Oil Microbiology. Also he has been granted 2 US patents on oil biotechnology. In addition, Prof. Rocha is actively engaged in providing professional consulting services to government agencies and private companies in matters related to the Sick Building Syndrome and oil bioremediation

ABSTRACT

Sick Building Syndrome (SBS) and Building Related Illness (BRI) are major concerns when considering the health of building occupants. SBS is defined by unspecific signs and symptoms that building occupants experience when they expose to combination of chemical, physical and biological factors (bioaerosols), whereas BRI is used to describe clinically identifiable illnesses due to specific airborne microorganisms. Because of this, the incorporation of biocides in construction materials is becoming more relevant. In this study we determined the type and density of airborne bacteria and fungi (bioaerosols) in two environments: surgery rooms and intensive care units from a public hospital (30 white areas) and office rooms from an electric company headquarter (27 non-white areas). Air samples were taken by the impaction method using a MAS-100 air sampler. Results showed that bacterial and fungal densities in the white areas ranged from 14 UFC/m^3 to 204 UFC/m^3 and from 2 UFC/m^3 to 204 UFC/m^3, respectively, whereas the results from the non-white areas depicted bacterial and fungal densities from 56 UFC/m^3 to 347 UFC/m^3 and from 6 UFC/m^3 to 800 UFC/m^3, respectively. According to the air microbial quality index used in this study 2 surgery rooms and 14 offices were found to be contaminated. Some bacterial pathogens were found in both buildings, whereas no pathogenic fungi were isolated.

Indoor air quality is a major concern when considering the health hazards related to the habitability of many buildings. The Sick Building Syndrome (SBS) is the accepted tern used to characterize unspecific signs and symptoms that building occupants experience when they expose to combination of chemical, physical and biological factors (bioaerosols) under which a range of severe health and comfort signs and symptoms appear to be correlated to the period of exposure in a "sick building" rather than to a particular illness or cause. Usually signs and symptoms tend to disappear as occupants abandon the sick building but return when they get back. However, the SBS may worsen and establish as a permanent clinical condition in the occupants (Oliva, 1992). According to the U. S. Environmental Protection Agency (EPA) signs and symptoms usually consists of headache, eye, nose or throat irritation, dry cough, dry or itchy skin dizziness and nausea, difficulty in concentrating, fatigue and sensitivity to odors among others (McGrath *et al.*, 1999).

On the other hand, The Building Related Illness (BRI) is the term used to describe clinically identifiable illnesses experienced by building occupants exposed to specific airborne bacteria or fungi atypically found in buildings. In cases related to both SBS and BRI, indoor air quality may be affected by bioaerosol constituents, such as bacteria and fungi, and by physical and chemical factors, which include mainly total organic compounds, temperature, humidity and breathable particles. Bioaerosols are key factors in the SBS. They usually contain dead and viable bacteria, fungi, parasites, protozoan, algae, virus and cell derivatives. Bioaerosols originate from dust particles and water drops aerosolized together with microorganisms and other biological particles (Mohr, 1997). Accordingly, it is imperative that physical-chemical factors, such as those mentioned above, as well as the right influx of fresh air must be controlled inside buildings to reduce the density of bioaerosols (Stetzenbach, 1997). Also, the nature of the building materials is especially relevant to bioaerosol generation, particularly those used in the construction of heating, ventilating and air conditioning systems (HVAC), carpets, walls, ceilings and other products and structures. In order to address this matter, the incorporation of biocides in the new generation of construction materials is becoming more relevant and will certainly aid in the prevention of microbial growth, and hence, in achieving a better indoor air quality (ISO14644-1,1999).

Bioaerosols determination is not only a way to characterize the indoor air quality but also a mean of testing the effectiveness and efficiency of biocides in building materials (Stewart *et al.*, 1995). Particularly, bacteria and fungi have proven to be the key bioaerosol constituents to be analyzed for indoor air quality assessment. To this effect several methods for bioaerosol trapping have been designed, which would allow the determination of microbial density as an indoor air quality index (Macher, 2000). Gravity, impregnation and impaction of bioaerosols on liquid and solid surfaces are the main physical principles upon which microbiological air sampler are base on. From these, the Impaction method is widely used for rapid and reliable microbial density determination.

On the other hand, the use of microbial (bacteria and fungi) density for evaluating indoor air quality has been confusing in the scientific literature, which has prevented the application of adequate legislation concerning occupational health standards. This could be explained on the basis that it has been difficult to correlate microbial density values with specific occupational health problems (Buttner *et al.*, 1997). However, several health organizations have proposed different criteria for assessing indoor air quality depending on the nature of the indoor area, mainly for white areas in hospital environments and non-white areas in regular indoor rooms. For non-white areas, The National Institute of Health and Occupational Wealth (NIOSH) has established that indoor areas with $\geq$ 1000 CFU/m^3 would be considered contaminated, on the other hand, The American Conference of Governmental Industrial Hygienists (ACGIH) has reduced value to 500 CFU/m^3 (Kalogerakis *et al.*, 2005). The Department of Health and Wealth of Canada also proposed that fungi density should not be higher than 50-500 CFU/m^3 depending on the fungi genera and fungi homogeneity The Department of Health of New York (NYHD) recognizes that security levels in terms of microbial density is hard to define, but indoor air evaluation should also considered microbial identification and outdoor air microbial density (Fabian *et al.*, 2005). The Spanish Association of Hospital Engineering (AEIH) has defined the following criteria for assessing indoor air quality in hospital environments (white areas): <10 CFU/m^3 *very clean*, 10-100 CFU/m^3 *clean*, 100-200 CFU/m^3 *acceptable* and > 200 CFU/m^3 *contaminated*. On the other hand, it has been stated that the indoor air microbial density should not be higher than outdoor air microbial density.

Based on all the above data and data obtained from research work done in many buildings in Venezuela, the Environmental Biotechnology Laboratory at the Universidad Simón Bolívar (LBA-USB) has established a regionalized indoor air microbial quality index taking into account the outdoor air microbial density. Also microbial identification is included for complete evaluation of the indoor air quality. In this study we evaluated the indoor air quality in surgery rooms and office rooms by the impaction method using the AEIH and the LBA-USB criteria as stated in materials and methods.

MATERIALS AND METHODS

Buildings

Two buildings were selected in this study due to their susceptibility to the SBS and BRI: a public general hospital (white areas) and an electrical building headquarter (non-white areas), both located in Caracas, Venezuela.

Air samples

Air samples were taken by the impaction method over solid surface using an Ardersen 1 platform type-microbiological air sampler MAS-100 (Merck, Germany). Indoor air samples from 33 surgery rooms and intensive care units (white areas) from the hospital and indoor air samples from 97 locations (non-white areas) distributed in 23 floors from the electrical building headquarter were analyzed for bacterial and fungal density. In both cases, microbial identification was undertaken for complete evaluation.

Viable bacteria and fungi densities were determined and used as an indoor air quality index. Microbial density was depicted as the mean value of Colony Forming Units in 1 m^3 air (CFU/m^3) from contaminated areas. Non-contaminated areas were omitted, unless they were relevant for discussion. In order to obtain representative air samples triplicates from 4 spots were taken from each location, all apart from each other.

Culture media for microbial growth.

Nutrient agar supplemented with a combination of amoxiciline and ampiciline was used to isolate bacteria, while discouraging fungal growth, whereas Sabouraud agar supplemented with itraconazole was used to allowed fungal growth, while inhibiting bacterial growth.

Incubation conditions for bacteria and fungi.

Samples were cultured at 30^0C for 6 days and viable colonies were count in a colony counter (Darkfield Quebec Colony Counter, Buffalo, NY, USA) to determine fungal and bacterial density.

Quantitative indoor air quality index for white areas.

In accordance with the AEIH, the following indoor air quality microbial index, applicable to white areas, was used:

<10 UFC/m^3 air: *very clean*
10-100 UFC/m^3: *clean*
100-200 UFC/m^3: *acceptable*
> 200 UFC/m^3: *contaminated*

Quantitative indoor air quality index for non-white areas

In accordance with both, the AEIH and the LBA-USB, the following indoor air quality microbial index, applicable to non-white areas, were used:

> 90 % below the outdoor air microbial density: *very clean*
50 -90% below the outdoor air microbial density: *clean*
0-50 % below the outdoor air microbial density: *acceptable*
0-50 % above the outdoor air microbial density: *low contamination*
50-99 % above the outdoor air microbial density: *medium contamination*
> 100% above the outdoor air microbial density: *high contamination*

Qualitative indoor air quality criteria for white and nonwhite areas

Bacteria and fungi isolates were identified by standard microbiological methods. The presence of pathogenic bacteria and fungi were used as the criteria for contamination.

RESULTS AND DISCUSSION

White areas in hospitals, particularly surgery rooms and intensive care units, must be regulated by strict cleanness standards. Cleanness protocols usually are applied to room infrastructure, medical and paramedical personal, surgery instruments, waste disposal and air. In this type of environment air may become a contaminant itself and in addition, can act as a vehicle to disseminate bioaerosols originated from many sources. In this study, air quality was analyzed using the bacterial and fungal density as an indoor air quality index. Microbial identification was also taken into account for a complete indoor air evaluation. The air in surgery rooms usually becomes a source of post-surgery infection diseases. The air in intensive care units, on the other hand, frequently offers a source of contamination while the patient is still institutionalized.

Air samples were taken from all surgery rooms and intensive care units including gynecology, urology, dermatology, oto-rhino, ophthalmology, neonatal, obstetrics, traumatology, cardiovascular; neurosurgery, general surgery and nephrology. The results indicated that bacterial density in most of the white areas ranged from 14 CFU/m^3 to 204 CFU/m^3, suggesting that air was either acceptable or clean according to the indoor air quality index used in this study. On the other hand, the M2 surgery room from the Oto-rhino service was the only white area that depicted bacterial contamination from the 33 white areas studied. Fig. 1 shows that the M2 room registered bacterial density values (202 CFU/m^3) higher than the reference upper limit value (200 CFU/m^3). Microbial identification from this surgery room revealed the isolation of *Acinetobacter junni* and *Stenotrophomonas maltophilia* (Table 1), two opportunistic bacterial pathogens. In addition, 8 pathogenic bacteria species were also isolated in the other white areas classified as acceptable o clean according to the indoor air quality index. It was interesting that some of the bacterial pathogens were found in the outdoor air, while others may have come from any other source. In contrast to the M2 surgery room from the Oto-rhino service, the neonatal surgery room depicted fungal contamination (204 CFU/m^3 air) and had not bacterial pathogens associated. These results were similar to those obtained in a neonatal surgery room at Kabala Hospital in Grace (Krajewska-Kulak *et al.*, 2007).

On the other hand, fungal density ranged from 2 CFU/m^3 to 204 CFU/m^3. Contrary to the 10 pathogenic species of bacteria found in the white areas (Table 1), pathogenic fungi were not found in the environments studied; however, 8 fungal genera and 5 fungal species were isolated in the white areas. *Penicillium* spp. (21 isolates), *Aspergillus niger* (15 isolates) and *Aspergillus fumigatus* (13 isolates) were the most frequent fungi found in this building (Table 2). Even though these fungi are non-pathogenic microorganisms, they have been associated with the SBS and with the BRI in immune-compromised building occupants (Reynolds *et al.*, 1990, Singh, 2005).

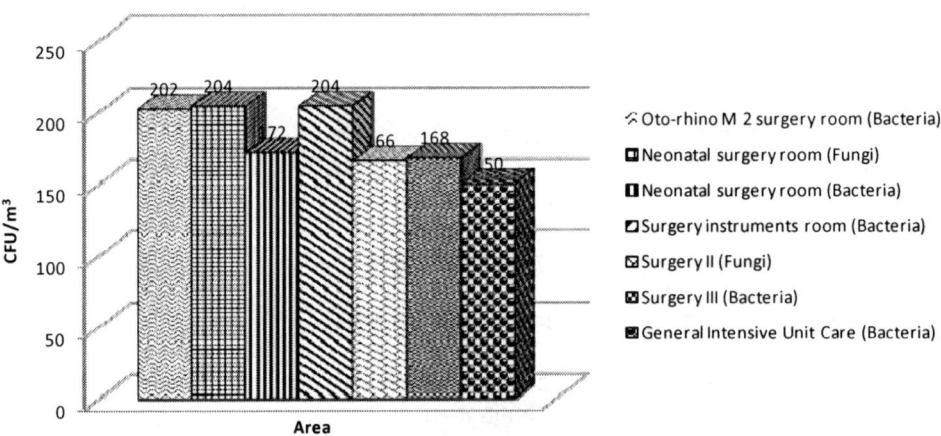

Fig. 1. Microbial (bacteria and fungi) density in white contaminated areas.

Area	Pathogen and opportunistic pathogen
Urology surgery	*Staphylococcus* coagulase-positive
Surgery I	*Staphylococcus saprophyticus*
Pediatric surgery (P room)	*Acinetobacter iwooffi*
General surgery (D room)	*Pseudomonas aeruginosa*
General surgery (C room)	*Acinetobacter iwooffi*
General surgery (G room)	*Acinetobacter baumannii*
General surgery /H room)	*Acinetobacter baumannii*
Cardiovascular surgery (A room)	*Acinetobacter junni*
Cardiovascular surgery (B room)	*Acinetobacter junni*
Traumatology surgery (K room)	*Staphylococcus saprophyticus*
Traumatology surgery (J room)	*Acinetobacter junni*
Hand traumatology surgery room	*Acinetobacter baumannii*
Intensive care unit (neuro-surgery)	*Pseudomonas aeruginosa*
Nefrology (kidney transplantation unit	*Serratina marcences*
Otorhino surgery (M2 room)	*Acinetobacter junni*
Otorhino surgery (M2 room)	*Stenotrophomonas maltophilia*

Table 1. Bacterial Pathogens in White Areas.

Fungal Isolate	Number of Isolates
Penicillium spp.	21
Aspergillus niger	15
Aspergillus fumigatus	13
Aspergillus flavus	4
Aspergillus clavatus	3
Scedosporium spp.	3
Fusarium spp.	3
Rhodotorula spp.	2
Paecilomyces spp.	1
Aspergillus terrus	1
Curvularia spp.	1
Trichodermo spp.	1

Table 2. Non-pathogenic Fungal Isolates from the White Areas.

  Contrary to the indoor air quality index in white areas, non-white areas in the office building also required the reference outdoor air microbial density value to determine this parameter (Kalageraskis *et al.*, 2005). As it was expected, all bacteria and fungi counts were well above the microbial density found in the hospital white areas. Around 14 locations showed microbial contamination. Bacterial contamination was detected in 10 locations with density values from 202 CFU/m^3 to 347 CFU/m^3 (Fig. 2). With the exception of Basement 4, the Restaurant (lunch room) and Floor 19, which depicted both bacterial and fungal contamination, the rest of the areas studied had one or the other type of microbial contamination. These results could be explained on the basis that Basement 4 was located in the deeper zone of the building with little influx of fresh air (Kreiss, 1993, Moritz *et al.*, 2001), the restaurant was exposed to food manipulation, which represented a significant source of bioaerosol generation, and Floor 19, where the all the electric system units were located, had a particular environment suitable for microbial growth. In contrast to the moderate number of pathogenic bacteria species found in the white areas, only two bacteria species (*Staphylococcus aureus* and *Staphylococcus saprophyticus*) were of clinical concern in the office building (Table 3). These pathogens were not found in the outdoor air samples, which would suggest that they may have been imported from other source and disseminated by the air HVAC system (Holt *et al.*, 1994). It is interesting to point out that *Escherichia coli* was isolated in Floor 5, which suggested a fecal source of bioaerosols (Fannin *et al.*, 1985).
  On the other hand, fungal contamination was registered in 8 locations. Fungi density varied from 6 CFU/m^3 to 800 CFU/m^3. Particularly, the microwave area, which is used for meal warming, was highly contaminated

(800 CFU/m^3). Also, the gym room, next to the lunch room (restaurant) depicted a very high fungi density. These results indicated that fungal contamination was more relevant in the office building than in the hospital, while bacterial contamination was higher in the hospital white areas than in the office building. This could be due to the predominant human bacterial infection in relation to fungal infection, biological waste disposal and the overcrowded concentration of sick occupants and visitors in the hospital, which would establish a microbial competition in favor of bacteria. In the office building, high humidity, low influx of outdoor fresh air and construction materials that favored fungal growth may have been the main reasons for this microbial distribution (Fisher and Dott, 2003, Gustarowsaka and Pioreowska, 2007). As in the hospital, *Penicillium* spp. and *Aspergillus* spp. were the predominant fungi found in this environment (McGrath *et al.*, 1999) and no pathogenic fungi were isolated. Particularly, 11 fungal genera and 3 fugal species were identified (Table 4).

Even though the determination of the possible sources of microbial contamination was not included in this study, preliminary observations revealed that the indoor air quality may have been affected by the air HVAC system. Both the hospital and the office building relayed on air HVAC systems to ventilate completely all the environments studied. Ductwork and the condensate water collection pans were important sources for bioaerosol generation, and hence, microbial contamination. Other sources for microbial contamination, such as old and inadequate carpets, file rooms, ceiling and wall materials, toilets, waste disposal, absence of HEPA filters and low influx of fresh air may have contributed to the overall increase of bacterial and fungal density (Al-Dagal and Fung, 1990, Heldman and Hedrick, 1971). Consequently, the incorporation of biocides in all types of indoor materials would aid in the health hazard preservation of building occupants. It should be noted there are just very few products in the market with biocide properties that can be incorporated on the fiber glass duct board, fiber glass lined ducts, ducts aluminum protective paints, walls, ceilings and floor materials (Bearg, 1993), which would be more effective and efficient than the controversial use of disinfectants, sanitizers or chemicals with antimicrobial effects. The implementation of biocides would, therefore, improve the cost-benefit ratio of indoor air cleaning and reduce the undesirable effects of adverse reactions commonly experienced by sensitive occupants and the intrinsic toxicity associated with the inadequate manipulation of chemicals.

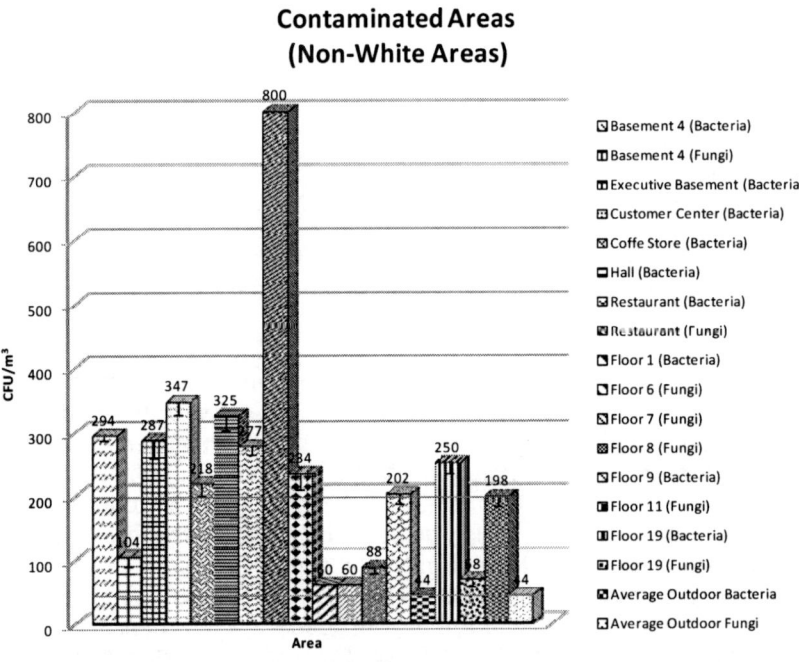

Fig. 2. Microbial (bacteria and fungi) density in non-white contaminated areas.
Red and blue lines indicate bacterial and fungal density from outdoor air.

Area	Pathogen and opportunistic pathogen
Basement 4	*Staphylococcus aureus*
Basement 2	*Staphylococcus aureus*
Executive Basement	*Staphylococcus aureus*
Customer Center	*Staphylococcus aureus*
Coffee Store	*Staphylococcus aureus* and *Staphylococcus saprophyticus*
Hall	*Staphylococcus aureus*
Restaurant (lunch room)	*Staphylococcus aureus* and *Staphylococcus saprophyticus*
Floor 3	*Staphylococcus saprophyticus* and *Staphylococcus aureus*
Floor 5	*E. coli*
Floor 7	*Staphylococcus aureus*
Floor 8	*Staphylococcus saprophyticus*
Floor 11	*Staphylococcus aureus* and *Staphylococcus saprophyticus*
Floor 14	*Staphylococcus saprophyticus*
Floor 15	*Staphylococcus saprophyticus*
Floor 16	*Staphylococcus saprophyticus*

Table 3. Bacterial Pathogens in Non-White Areas

Fungal Isolates	Number of Isolates
Penicillium spp.	21
Aspergillus níger	19
Aspergillus flavus	8
Aspergillus spp.	7
Rhodotorula spp.	6
Aspergillus fumigatus	4
Curvularia spp.	4
Cladosporium spp.	4
Rhizopus spp.	2
Scedosporium spp.	2
Fusarium spp.	2
Alternaria spp.	1
Nigrospora spp.	1
Paecilomyces spp.	1

Table 4. Fungal Isolates from the Non-White Areas

All the data shown in this study indicated that the indoor air microbial quality index can be a very convenient and reliable way to evaluate the SBS, BRI (Kalogeraskis *et al.*, 2005) and the effectiveness and efficiency of biocides incorporated in new construction materials.

BIBLIOGRAPHY

1. Al-Dagal, M and Fung, D. 1990. Aeromicrobiology: A Review. Food Science and Nutrition. 29 (5):333-340.

2. Bearg, D. 1993. Indoor Air Quality and HVAC systems. Lewis Publishing, USA.

3. Buttner, M., Willeke, K.,Grinshpun, S. 1997. Sampling and Analysis of Airborne Microorganisms. Pag.: 629-640 In Hurst, C., Knudsen, G., McInerney, M., Stetzenbach, L., Walter M. Manual of Environmental Microbiology. American Society for Microbiology. USA.

4. Fabin, M., Miller, S., Reponen, T., Hernández, M. 2005. Ambient Bioaerosol Indices for Indoor Air Quality Assessments of Flood Reclamation. Journal of Aerosol Science. 36:763-783.

5. Fischer, G. Y. and Dott, W. 2003. Relevance of Airborne Fungi and their Secondary Metabolites for Environmental, Occupational and Indoor Hygiene. Archives of Microbiology. 179:75-82.

6. Gustarowsaka, B and Pioreowska, M. 2007. Methods of Mycological Analysis in Buildings. Building and Environment. 42:1843-1850.

7. Heldman, D. and Hedrick, T. 1971. Airborne Contamination Control in Food Processing Plants. Michian State Agriculture Experimental Station Research Bull. 33:1

8. ISO14644-1,1999. Cleanrooms and Associated Controller Environments. Part I-Classification of air cleanliness. CEN- European Commission for standardization.

9. Krajewska-Kulak, E., Lukaszuk, C., Tsokantaridis, Ch., Hatzopoulu, A., Theodosopoyly, E., Hatzmanasi, D., Kosmois, D. 2007. Indoor Air Studies of Fungi Contamination at the Neonatal Department and Intensive Care Unit and Palliative Care in Kavala Hospital in Grace. Advances in Medical Sciences. 52 (1):12-14.

10. Kalogeraskis, N., Paschli, D., Lekaditis, V., Pandidou, A., Eleftheriadis, K., Lazaridis, M. 2005. Indoor Air Quality-Bioaerosols Measurement in Domestic and Office Premises. Aerosol Science. 36:751-761.

11. Macher, J. 2000. Air Sampling Methods for Biological Contaminants. http://www.thermoandersen.com/Macher.htm

12. McGrath, J., Wong, W., Cooley, J., Straus, D. 1999. Continually Measured Fungal Profiles in Sick Building Syndrome. Current Microbiology. 38:33-36.

13. Mohr, A. 1997. Fate and Transport of Microorganisms in Air, pag.: 641-650 In Hurst, C., Knudsen, G., McInerney, M., Stetzenbach, L., Walter M. Manual of Environmental Microbiology. American Society for Microbiology. USA.

14. Singh, J. 2005. Toxic Moulds and Indoor Air Quality. Indoor and Built Environment. 14 (3-4):229-234.

15. Oliva, M. 1992. Síndrome del Edificio Enfermo ¿Trabaja Usted en la Oficina Siniestra?. Muy Interesante. 6 (76): 12-19.

16. Reynolds, S., Streifel, A., McJilton, C. 1990. Elevated Airborne Concentration of Fungi in Residential and Office Environments. American Industrial Hygiene Association Journal. 51 (11):601-604.

17. Stezenbach, L. 1997. Introduction to Aerobiology, pag.: 619-628 In Hurst, C., Knudsen, G., McInerney, M., Stetzenbach, L., Walter M. Manual of Environmental Microbiology. American Society for Microbiology. USA.

18. Stewart, S. L., Grinshpun, S. A., Willeke, K., Terzieva, S., Ulevicius, V., Donnelly, J. 1995. Effect of Impact Stress on Microbial Recovery on an Agar Surface. Applied and Environmental Microbiology. 61 (4):1232-1239.

19. Thorne, P. S., Kiekhaefer, M. S., Whitten, P., Donham, K. J. 1992. Comparison Bioaerosol Sampling Methods in Barns Housing Swines. Applied and Environmental Microbiology. 58 (8): 2543-2551.

20. Witarnen, G., Miettinen, H., Pahkala, S., Enbom, S., Vanne, L., 2002. Clean Air Solutions in Food Processing. Espoo. VTT Publications 482. Pag.: 95.

FUNGAL COLONISATION AND CONTAMINATION OF CINEMATOGRAPHIC FILM: IMPLICATIONS FOR FILM AND ARCHIVISTS

Gavin Bingley
Manchester Metropolitan University
John Dalton Building, Chester Street, Manchester, M1 5GD, UK
Email: gavdb@hotmail.co.uk

BIOGRAPHICAL NOTE

Gavin Bingley is just about to start the second year of his PhD at Manchester Metropolitan University after graduating from his first degree in microbiology in 2008. After studying mainly bacteriology at undergraduate level, Gavin has moved into mycology, specialising on the fungal contamination of cinematographic film due to interest from the North West Film Archive in Manchester (part of Manchester Metropolitan University) and the British Film Institute (BFI) based in Berkhamsted. Gavin is a member of both the Society For General Microbiology and The International Biodeterioration and Biodegradation Society.

ABSTRACT

Cinematographic film is composed of 3 generic layers: a polymer based base support, a photosensitive emulsion coating and a binder based on gelatine. Studies were carried out on 19 film reels using air sampling, to quantify spores released from contaminated film during a simulated inspection process, in order to assess exposure of archivists to spores. Organisms present were identified and screened for gelatinase production, since gelatine is the major substrate for fungal growth on the film. The majority of fungi present were *Aspergillus* and *Penicillium* species, 16 out of 30 isolates of which produced gelatinase. For some films, released spore numbers exceeded the recommended safe exposure levels of $1000cfu/m^3$. Some films appeared contaminated, but no fungal growth was detected post-inspection. However, hyphal growth was evident across film frames, indicating that the damage may have taken place in the past. This study indicated a need for detection of fungal contamination of film, the presence of viable fungal spores, and safe handling recommendations for film archivists.

Slide 1

Fungal Colonisation And Contamination Of Cinematographic Film: Implications For Film And Archivists

Gavin Bingley

Manchester
Metropolitan
University

Professor Joanna Verran
Dr Gordon Craig
Mark Bodner (NWFA)
Dr Ina Stephan (BAM, Berlin)

Slide 2

Introduction

2 main types of deterioration in archive materials:

. **Chemical** – breakdown of material caused by exposure to heat or moisture.

. **Biological** –Physical growth of microorganisms or production of metabolites such as enzymes or acids.

Materials in Film

Film

• **3 layers**
Base (3 types): Cellulose Nitrate, Cellulose Acetate and Polyethylene

• Photosensitive emulsion coating (dyes in colour, metallic silver in black and white),

• Binding agent which is based on gelatine.

Slide 3

Archives

• Conserve Materials – Copy, Clean

• Reduce Further damage – e.g. prevent microbial growth

• Store

North West Film Archive (NWFA)

Slide 4

Importance of Materials Kept in Archives

- Cultural Importance – History, Society

- Personal Value – Home Movies and Photographs

Problems

- Deterioration of film

- Storage of the film – conditions and cost of maintenance

- Health risks to archivists?

Slide 5

Aims

To identify the microorganisms responsible for spoilage and contamination of film and to provide recommendations for safe handling of such archive materials.

Film Reel Presenting Heavy Fungal Contamination
(Mark Bodner (NWFA)

Slide 6

Objectives

1) To determine the number of fungal spores released during the inspection of cinematographic film.

2) To identify the fungi which cause biodeterioration of cinematographic film and paper.

3) To detect the microbial enzymes responsible for biodeterioration, in order to define key deteriogens

4) To assess the risk posed to archivists during handling of paper and film heritage.

Slide 7

Simulated Film Inspection

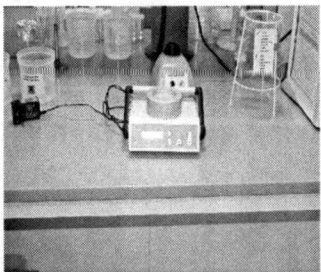

GS100 Air Sample (Desaga)

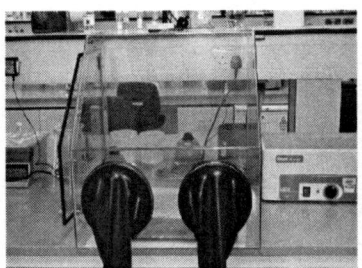

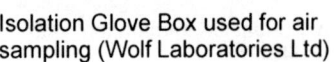

Isolation Glove Box used for air sampling (Wolf Laboratories Ltd)

Slide 8

Typical Air Sampling

Malt Extract Agar
plate with colonies
isolated from air
sampling.

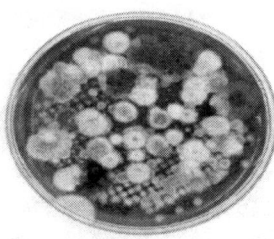

• Relative counts were based on colony morphology

•Which isolates were primarily responsible for film
deterioration?

•Which isolates were contaminants and which were
colonisers?

Slide 9

Fungi Isolated

Domestic Origin:

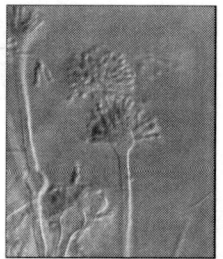

- *Penicillium sp.*

- *Alternaria sp.*

- *Cladosporium sp.*

- *Aspergillus sp.*

- Soil Fungi – *Mucor, Trichoderma*

Of particular interest (health):

- *Aspergillus niger, Aspergillus fumigatus*

- *Stachybotrys chartarum*

Slide 10

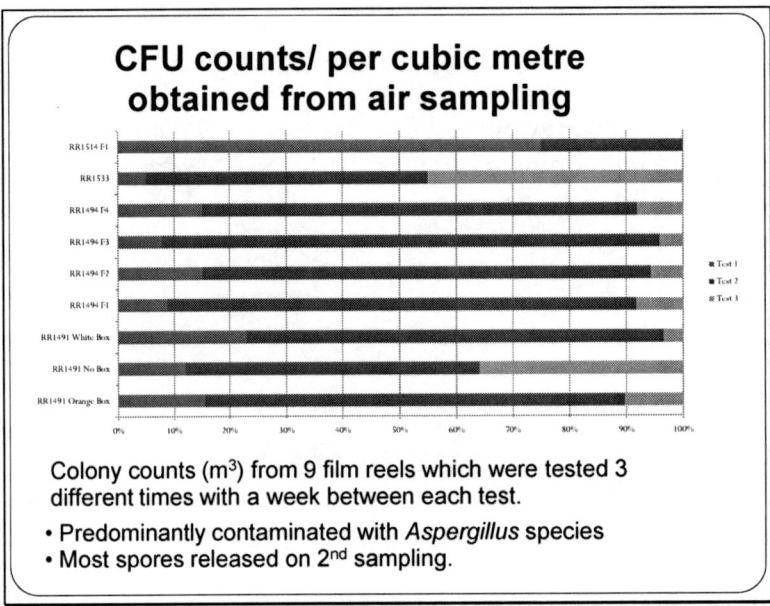

Slide 11

Isolate Frequencies based on visualisation of colony morphologies

Isolate Name	Test 1	Test 2	Test 3
RR1399 I1	920 +/- 511.18	1000 (N=1)	2140 +/- 608.28
RR1399 I2	100 +/- 65.67	280 (N=1)	273.33 +/- 61.1
RR1399 I3	3623 +/- 80.21	4240 (N=1)	253.33 +/- 102.63
RR1399 I4	10 +/- 10	0 (N=1)	440 +/- 124.9
RR1399 I5	23.33 +/- 20.82	10 (N=1)	120 +/- 20
RR1399 I6	396.67 +/- 20.82	220 (N=1)	386.67 +/- 61.1

Sample showing frequencies of isolation of different moulds from reel RR1399, where colonies were presumed to be the same based on morphology.
• Heaviest contaminants were *Aspergillus* species.

Slide 12

Gelatinase Production

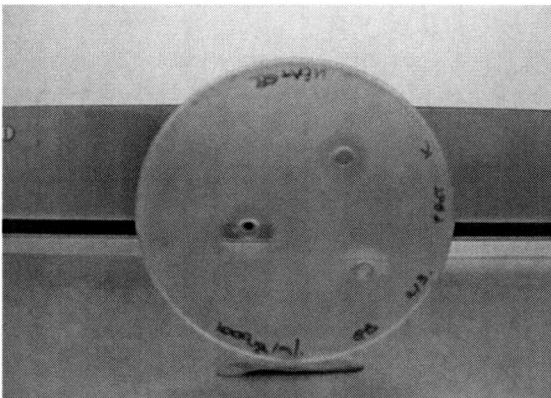

Fig 5. MEA and gelatine agar plate with clear zones produced by gelatinase enzymes isolated from the supernatants of fungi grown in broth culture

Slide 13

Testing For Gelatinase Production

Isolate Name	Genus	Gelatinase on Agar +/-	Liquidation of Supernatant	Size of Clear zone (mm) if applicable	Equivalent Proteinase K concentration (ug/ml)
A. versicolor	Aspergillus	Positive	Positive	4	20
RR1491 OB1	Aspergillus	Positive	Positive	3.5	10
RR1491 NB6	Aspergillus	Positive	Positive	4	20
RR1514 F1 I1	Penicillium	Positive	Positive	2	1.25
JWP1 I1	Alternaria	Negative	N/A	N/A	N/A

Results of gelatinase assays from fungi isolated from cinematographic film.

- Zone of enzyme activity

- If clear zone is present grow in broth containing MEB and gelatine and assay the supernatant. Zones compared to clear zones produced by known concentrations of Proteinase K (Sigma)

Slide 14

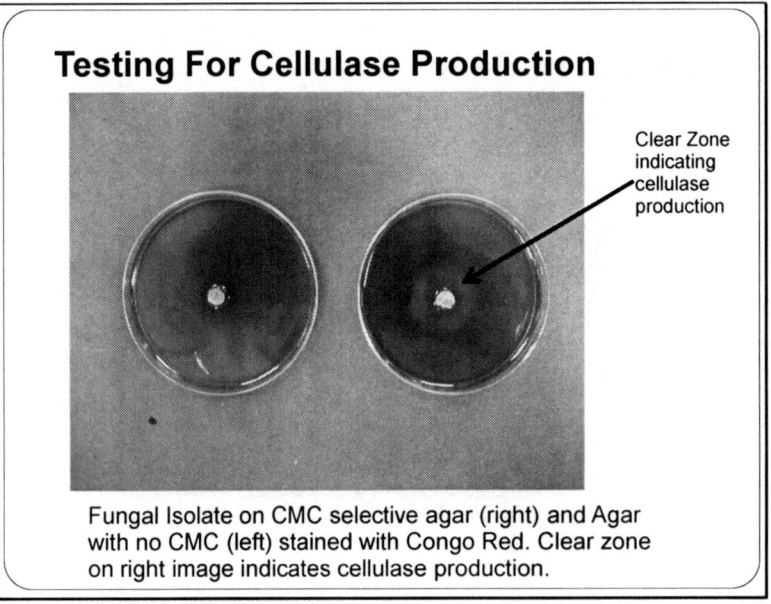

Testing For Cellulase Production

Clear Zone indicating cellulase production

Fungal Isolate on CMC selective agar (right) and Agar with no CMC (left) stained with Congo Red. Clear zone on right image indicates cellulase production.

Slide 15

Fungi on Film Slide As Seen Through A Projector

Image Capture of Film Slide Showing Fungal Mycelial Growth On The Face And Right Shoulder

Slide 16

Assessing Fungal Growth On Film Strips

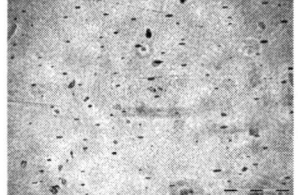

Clean Test Film (x10 magnification)

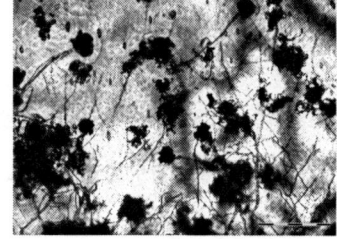

A. versicolor growing on inoculated test film (x10 magnification)

Fungi Seen Growing On A Film Frame Under a Microscope (x10). 'FI' from the word 'FILM' is shown here.

Slide 17

A Role For Biocides?

Current Methods of Removing Fungi From Film

Physical

• Brush to remove mould on the surface of the roll – if mould is dry

• If thick and vegetative growth – rejected/ discarded/stored

Chemical

• Killed using Paraformaldehyde and Bifenyl – Bag Inserted into each film can (Vietnam)

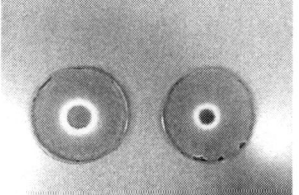

Indirect effect of Essential Oil Vapour On *Penicillium chrysogenum*

Slide 18

Conclusions

• Many fungi involved in biodeterioration of film – predominantly *Aspergillus* and *Penicillium* species

• Inspection of some films released spore numbers above levels deemed 'safe'.

•Spoilage of film even if mould is 'dead' or unviable

• Utilisation of substrates suggests capability of biodeterioration.

• Fungal growth can distort images
Future Work

• Advice To Archivists (Decision Tree)

• Detection Kit for Microbial Volatile Organic Compounds (MVOCs)

•Additional Archives/ materials e.g videotapes

THE USE OF NOVEL, NANO-LIME DISPERSIONS TO MODIFY THE SUSCEPTIBILITY OF MATERIALS TO MICROBIAL GROWTH

Gerald Ziegenbalg[1], Peter D Askew[2]

[1] IBZ-Salzchemie GmbH & Co.KG

Halsbrücker Strasse 34, 09599 Freiberg, Germany

Tel: +49 (0) 3731 200 155 Fax: +49 (0) 3731 200 156   gerald.ziegenbalg@ibz-freiberg.de

[2] IMSL, Pale Lane, Hartley Wintney, Hants, RG27 8DH, UK

Tel: +44 (0) 1252 627676 Fax: +44 (0) 1252 627678   peter.askew@imsl-uk.com

BIOGRAPHICAL NOTE

Gerald Ziegenbalg

Founder and CEO of IBZ-Salzchemie GmbH & Co.KG

1998:	Dr. habil. (Habilitation, Chemical Engineering),
1990:	Dr. rer. nat. (PhD)
1986	Dipl.-Chem. (M.Sc., Inorganic chemistry)

ABSTRACT

Biodeterioration is a serious problem not only in the protection of cultural heritage but in many cases also in the refurbishment of buildings. The growth of fungi and algae can result in the damage of natural and artificial stone and can cause health problems and allergies. Sols containing stable dispersed nano-lime particles were developed in order to strengthen degraded stone and mortars and to remove microbiological growth in an eco-friendly way, without the use of chlorine or quaternary ammonium compounds. Ethanolic suspensions of lime ($Ca(OH)_2$) nano-particles are used. They contain lime particles having sizes in the range between 50 and 250 nm. Safe removal of biological growth is achieved by the dehydrating action of ethanol in combination with the creation of alkaline conditions by the lime particles. The small size of the lime particles guarantees deep penetration into stone, mortar and plaster structures. This article gives an overview about the properties of calcium hydroxide nano-dispersions and summarises field tests in the ancient theatre of Megalopolis (Greece) to remove the growth of lichens.

1   Introduction

Lime ($Ca(OH)_2$) and limestone in all its variations have found manifold applications in all construction areas for centuries. Lime based mortars, plaster or stuccos as well as lime wash or lime paints are materials known for producing a healthy environment while allowing the creation of aesthetically pleasing surfaces and structures. Traditional lime-based materials are not only important for the conservation of historic structures, buildings and monuments but they have also enjoyed a renaissance in construction as eco-friendly materials.

For the conservation of historic mortars and plasters, limestone and marble especially as well as for stucco and wall paintings, materials are necessary which are fully compatible to those used during the original construction. Lime plays an important rule in that. Standard lime slurries are, however, not able to penetrate into small fissures, cracks or pores due to the large size of the suspended particles.
Sols containing stable dispersed nano-lime particles were developed to overcome this obstacle. The particles are stable when dispersed in different alcohols. After evaporation of the solvent, solid calcium hydroxide particles are deposited in the treated materials. These react with atmospheric carbon dioxide forming calcium carbonate in a similar manner to conventional calcium hydroxide suspensions.

It is well known that microbiogical growth contributes not only to the aesthetic defacement of structural surfaces but can also result in damage to the materials and structures themselves. Fungi and algae colonise many natural and man-made materials used in the construction of buildings and monuments and produce disfiguring stains (mildew) and structural deterioration. Furthermore, their presence can increase the moisture holding capacity of materials and thereby exacerbate damage associated with moisture by extending periods of wetness. The majority of remediation actions for biological growth are based on biocide / disinfectant treatments with materials such as sodium hypochlorite (or chlorine donors) and quaternary ammonium compounds being commonly employed. These materials are relatively hazardous to handle and can have deleterious effects on substrates and, in large quantity, be damaging to the environment. In addition, removal of growth is often achieved by scrubbing / application of water under high pressure (often with no biocidal treatment) and this can lead to the generation of aerosols which can in themselves present hazards to health as well as cause abrasion damage to the surface of the materials being cleaned.

Comprehensive tests have shown that the application of ethanolic nano-lime sols for the consolidation of mortars is accompanied by safe removal of biological growth. This article, which is based on investigations carried out within the EU Framework VII STONECORE project, describes the concept of using $Ca(OH)_2$ - nano-paticles as a consolidant and agent for the removal of fungi and algae from surfaces and summarises preliminary studies into its use as a modifier of synthetic materials intended to demonstrate antimicrobial properties.

2 Characteristics of colloidal $Ca(OH)_2$ sols

Conventional lime hydrate suspensions are characterised by particles having sizes in the µm range. A typical particle size distribution is given, in comparison with nano-lime, in Fig. 1. The extremely fine size of synthetic nano-lime results from its preparation, which is based on chemical synthesis. The particles are stable when suspended in either ethanol, iso-propanol or n-propanol. Typical concentrations are in the range between 5 and 75 g/L. Due to the low particle size, stable sols are formed that means the solids do not sediment for a long time. The nano-lime suspensions are white to opal liquids. The small size of the calcium hydroxide particles guarantees a deep penetration into deteriorated zones (Fig. 2).

The penetration depth depends on many parameters. The more important of which are:
- Mineralogical composition and surface properties.
- Porosity and absorbency.
- Moisture content.
- Concentration of the nano-lime
- Used solvent.
- Temperature and air humidity.

All solvents evaporate without leaving any residues. Components which could deteriorate stone, mortar or plaster are not formed.

The stability of the sols is caused by electrostatic repulsion. In alcoholic solvents, calcium hydroxide particles have a positive surface charge. As long as these remain stable, the sols do not settle and a shelf life of between thee and five months is possible.

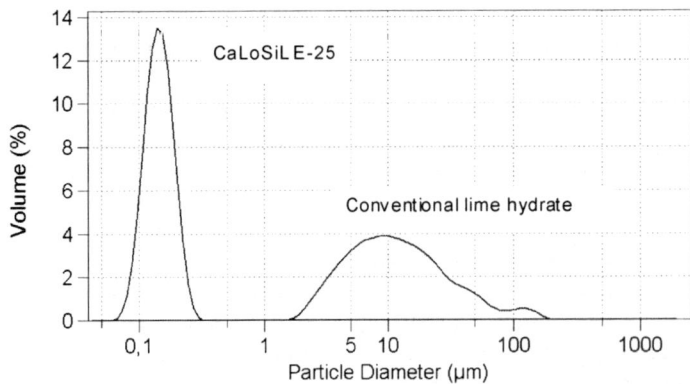

Fig. 1: Particle size distribution of CaLoSiL E-25 in comparison to traditional lime hydrate

Fig. 2: Porous sandstone absorbing CaLoSiL E-25 by capillary action

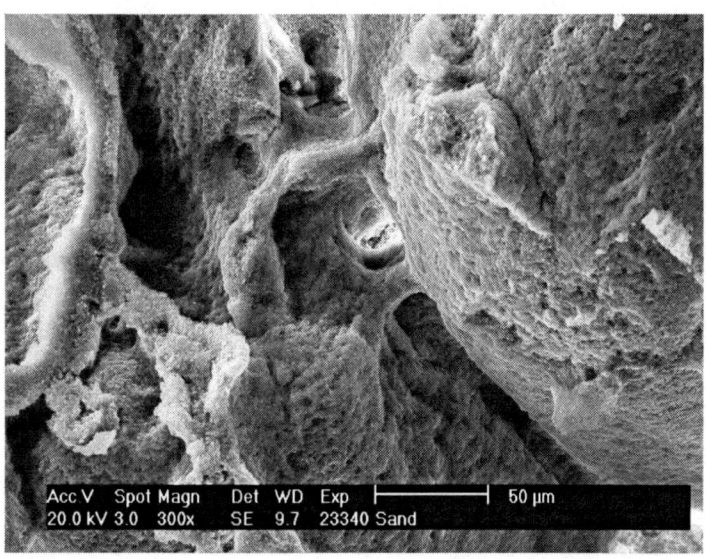

Fig. 3: Fine structure of $CaCO_3$ crystals in matrix connecting sand grains

Products based on nano-lime are commercially available under the trade name "CaLoSiL®". IBZ-Salzchemie GmbH &Co.KG (Germany) is the producer. The letters behind the name indicate the used solvent used, the numbers give the total calcium hydroxide concentration in g/L. For example, E-25 means, 25 g/L calcium hydroxide dispersed in ethanol. CaLoSiL® is applied to stone/plaster in a flow coating, dipping and/or injection procedure.

Typical $CaCO_3$ nano-crystals formed after carbonation are shown in Fig. 3. Examples of successful application of CaLoSiL® are the medieval cellars of the Middle Castle in Malbork (Poland), the facade of the Church of the Visitation Order in Warsaw (Poland), the Lichfield Angle (UK) or the Capilla General de Animas, (Santiago de Compostela, Spain) and the Xanten Cathedral (Germany).

3   Experimental

Treatment of ancient limestone
The STONECORE project, which is funded within the FP-7 programme of the European Commission, combines fundamental and applied research. The materials developed in the laboratory are tested on real objects under different climatic conditions. In this context, the ancient theatre in Megalopolis (Greece) was selected for testing colloidal lime as a consolidant as well as a material for stopping biological growth.

The tests were divided into the steps:

- recover, isolate and identify the microorganisms,
- treat with different biocides and nano-lime suspensions
- characterise the effect of the treatment
- long term survey

Once located and identified photographically, the test area was sampled using a sterile swab moistened with sterile distilled water. To help differentiate isolates that were simply surviving on the surfaces from those that were colonising the surface an adhesive tape imprint was taken from an area adjacent to where the swab sample was taken.

Swabbing method
Sterile cotton swabs moistened with sterile distilled water were used to remove growth from an area of 3 x 3 cm at selected locations at each site / object. Where necessary, more than one swab was used to remove as much growth as possible such that a surface bare of growth was revealed. The swabs were then placed into individual sterile containers (30 ml) per area (*ie* multiple swabs were combined) and then transported to the laboratory with minimum delay for isolation and characterisation.

Tape method
The swab recovery method will recover both growing and dormant cells from a surface. In many cases, the observations and photographic record made at the site will help to differentiate true growth from incidental contamination. To complement this approach, adhesive tape imprints can be employed to show the pattern of growth present on the surface at the time of sampling. A short (5 cm) section of clear adhesive tape was placed onto the surface in an area adjacent to where the swab sampling had been performed (and which demonstrated similar growth) and was pressed firmly onto the surface by applying pressure by hand.

Isolation of Fungi
- Swabs were streaked onto multiple plates of Saboraud Dextrose Agar + Chloramphenicol (SDA) and then incubated at 20°C for up 10 days.
- When suitable growth was observed, individual isolates were transferred using a loop / scalpel onto fresh plates of SDA, Malt Extract Agar (MEA) *etc*.
- This was repeated until a pure sub-culture was obtained
- The pure isolates were then transferred to Malt Extract Agar slopes and, when grown, these were stored at 4°C for use in the growth study phase of the project.

Isolation of Algae
- After streaking onto SDA the swab was placed into Jaworski Medium and incubated at 20°C under light (1.8 Klux, 16 hour photo-period) for up to 6 weeks
- Any growth was purified by separation and sub-culturing into fresh Jaworski Medium fortified with carbendazim (200 mg l-1) and incubated at 20°C under light (1.8 Klux, 16 hour photo-period) for up to 6 weeks. This was repeated until axenic cultures had been obtained.

4   Results

An area on the first seating row of the right hand isle of the theatre was selected for sampling and testing (Fig. 4). The microbiology of the exposed area is dominated by lichens (Fig. 5). Despite this, eight distinct fungal strains including again *Penecillium brevicompactum*, were isolated although several other strains producing 'sterile' mycelium were obtained (it is considered likely that they may prove to be the mycombiont component of the lichens present). *Protocooccus sp* was present along with a small green unicellular algae probably of the genus *Chlorella*. In addition, two unicellular blue green algae were isolated (probably *Cyanothece sp* and *Gloeocapsa sp*) as well as a filamentous blue green species (probably *Phormidium sp*).

Fig. 4: The ancient theatre in Megalopolis (Greece)

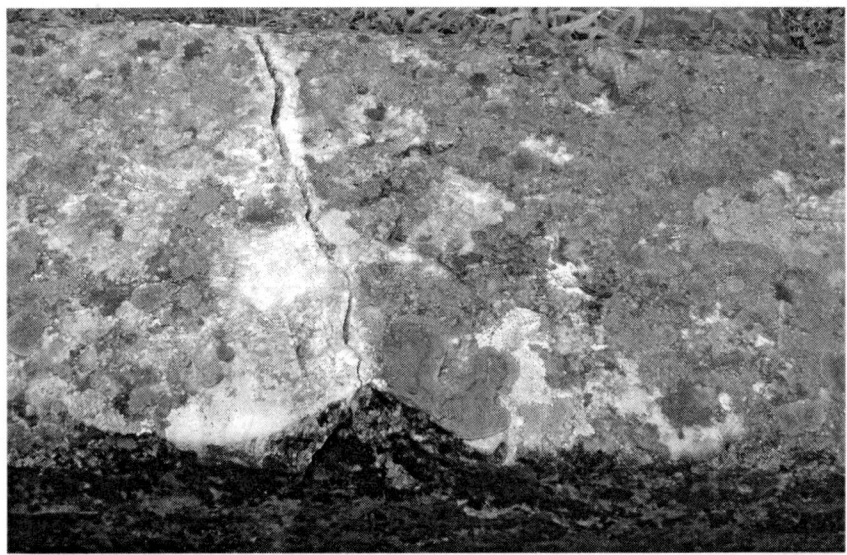

Fig. 5: Test area: Dense growth of lichens around the crack

Fig. 6: After the application of CaLoSiL

Tab. 1: Results of the treatment with different biocides

Sample number		Fungal growth at 9 days	Fungal growth at 16 days	Fungal growth at 1 month
1	Cleaning with H_2O_2	White mucoid growth (some possible Penicillium)	White mucoid growth	Mucoid growth
2	Physically cleaning	Grey fungi (some possible Penicillium)	Grey fungi (some possible Penicillium)	Brown /grey fungus (some possible Penicillium)
3	Washing with iso-propanol	Off grey fungi (Some possible Penicillium)	Off grey fungi (Some possible Penicillium	Brown fungus
4	Cleaning with H_2O_2 and two treatments with nano-lime (CaLoSiL® E-25)	No growth	No growth	No growth
5	Physically cleaning and two treatments with nano-lime (CaLoSiL® E-25)	No growth	Possible Penicillium	Possible Penicillium
6	Treatment with nano-lime (CaLoSiL® E-25), no cleaning	No growth	Green /brown growth	Green / brown growth

The following treatment procedures were tested in order to remove the lichens:

- Cleaning with H_2O_2
- Physically cleaning
- Washing with iso-propanol
- Cleaning with H_2O_2 and two treatments with nano-lime (CaLoSiL® E-25)
- Physically cleaning and two treatments with nano-lime (CaLoSiL® E-25)
- Treatment with nano-lime (CaLoSiL® E-25), no cleaning

At 24 hours after the application of the different biocides, the test areas (Fig. 6) were sampled. The effect of the treatment was determined by standard growth tests in the laboratory as described above. The results are given in Tab. 1. Physical cleaning as well as treatment with H_2O_2 or iso-propanol was not sufficient to remove all of the biological growth and the use of nano-lime only proved more efficient. However, complete prevention of new fungal growth could not be achieved by a single treatment. As shown in Fig. 5 lichens form very dense layers on the surface of the stone. It is considered probable that the nano-lime dispersions were not capable of penetrating the lichens completely. The treatment was performed in July at temperatures of around 30 °C resulting in a relative fast evaporation of the ethanol. It is assumed that the retention time of the ethanol on the surface of the stone was not sufficient to penetrate and kill the the biological growth. This may also be the explanation why the treatment with iso-propanol was ineffective. The combined application of H_2O_2 and nano-lime resulted in complete disinfection and fresh biological growth could not be detected one month after treatment suggesting that the modified surface is proving more resistant to colonisation that the surrounding weathered stone. The survey is still continuing.

The stones in the ancient theatre in Megalopolis are characterised by micro-and macro cracks. While nano-fissures could be filled with CaLoSiL® E-25 directly, larger fractures were treated with special injection grouts containing nano-lime as binder. In both cases a good connection to the stone surfaces was found. The injection grouts dried without any shrinkage guaranteeing a long time stabilisation.

Modification of synthetic polymers with nano lime dispersions

Samples of unfortified polypropylene (PP) and acrylonitrile butadiene styrene (ABS) were treated by dipping into either CaLoSiL® E-25 or nano particle dispersed $BaCO_3$ and then allowed to dry. The treated materials were then assessed for antibacterial properties using ISO 22196 using *Escherichia coli* and methicillin resistant *Staphylococcus aureus* (MRSA).

The results are shown in Figures 7 and below.

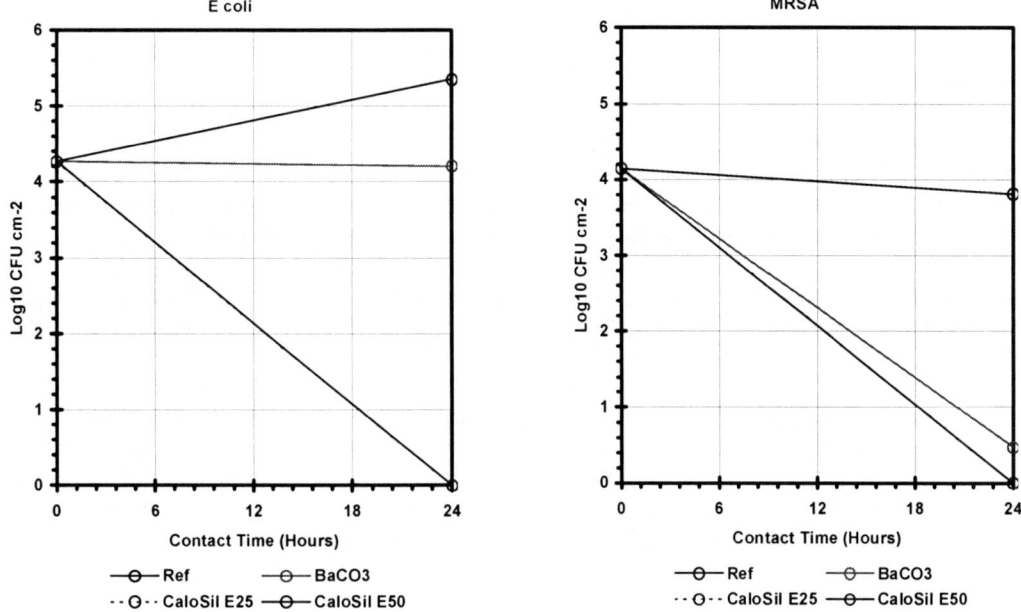

Figure 7: Effect of treatments on survival of *E coli* and MRSA on PP

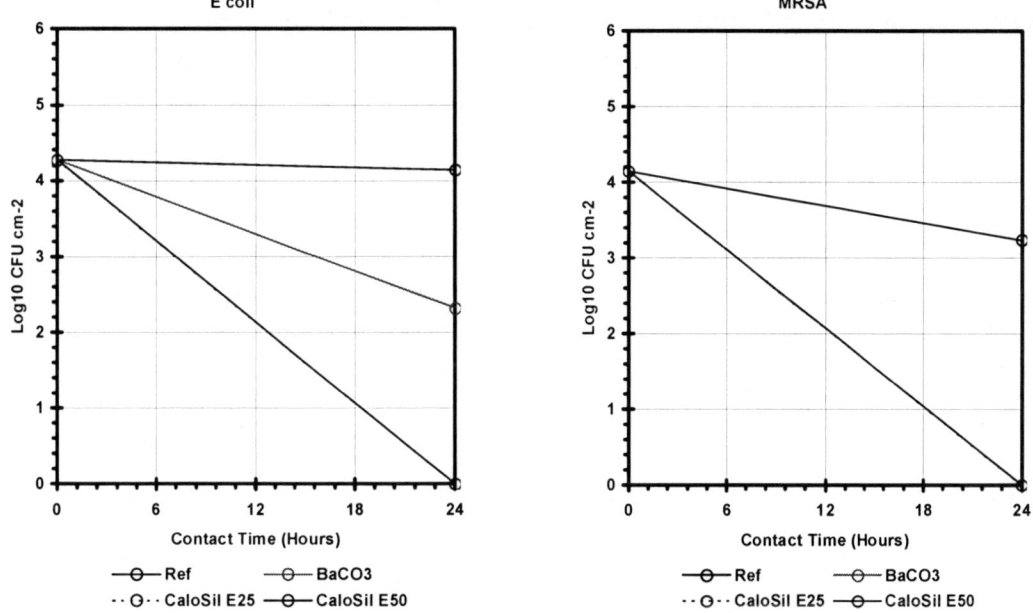

Figure 8: Effect of treatments on survival of *E coli* and MRSA on ABS

It can be seen from the figures that the populations of both *E coli* and MRSA remained viable when held in contact with the surface of untreated PP and ABS. In contrast, both populations were reduced to below the limit of detection after being held in contact with surface on which nano-lime had been deposited. Growth of *E coli* was inhibited by the presence of $BaCO_3$ on PP and the effect on the survival was significantly less on ABS that that observed with nano-lime. Greater activity was however, observed with MRSA.

5   Conclusions

Biological growth is capable of deteriorating and damaging natural and artificial stone, especially where the stone is exposed to conditions of high humidity. Treatment with calcium hydroxide nano-sols combines the following advantages:

- Ethanol is an extremely effective disinfectant acting as a dehydrating agent damaging the cell membranes of algae and fungi and causing the denaturation of intracellular proteins and proteins associated with cell walls. This can result in the stopping of microbiological activity in treated areas.
- Lime creates alkaline conditions under which new biological growth is inhibited.

The removal of biological growth from stone and mortar is combined with structural consolidation of the treated areas because nano-lime is converted by reaction with atmospheric carbon dioxide with time into calcium carbonate.

The antimicrobial properties observed following the treatment of stone can also be reproduced on synthetic materials by the deposition of nano-lime on their surface. Effects are less pronounced with $BaCO_3$. The resulting coatings are not likely to prove highly durable but do demonstrate that antibacterial properties can be achieved by such treatments. Studies are ongoing to apply such treatments to other synthetic and modified natural materials in which durability is anticipated to be significantly enhanced.

6   Acknowledgements

The research leading to these results has received funding from the European Community's Seventh Framework Programme [FP7/2007-2013] under grant agreement No 213651 (STONECORE).

DESIGNING NOVEL ANTIMICROBIAL SURFACES, POLYMERS AND HYDROGELS

Ayusman Sen

Department of Chemistry, The Pennsylvania State University, University Park, Pennsylvania 16802, USA
Phone: (814) 863-2460; FAX: (814) 863-5319 E-mail: asen@psu.edu
http://research.chem.psu.edu/axsgroup/ http://network.nature.com/people/catalyst/blog

BIOGRAPHICAL NOTE

PERSONAL	- Born: January 5, 1951; Calcutta, India
EDUCATION	- 1970 B.Sc. (Honours), University of Calcutta, India
	1973 M. Sc., Indian Institute of Technology, Kanpur, India
	1978 Ph. D., University of Chicago, USA

PROFESSIONAL

7/2004 - 6/2009	Head, Department of Chemistry, Penn State University
1/2010 - present	Distinguished Professor of Chemistry, Penn State University
7/1989 - 12/2009	Professor of Chemistry, Penn State University
7/1984 - 6/1989	Associate Professor of Chemistry, Penn State University
9/1979 - 6/1984	Assistant Professor of Chemistry, Penn State University
7/1978 - 6/1979	Research Fellow, California Institute of Techno logy

SELECTED HONORS AND FELLOWSHIPS

1967 - 74	National Science Talent Search Scholarship, Government of India
1982 - 84	Young Investigator Award, Chevron Research Company
1984 - 88	Alfred P. Sloan Research Fellow
1987 - 88	Paul J. Flory Sabbatical Award, IBM
1993	Imperial Oil Distinguished Lecturer, University of Toronto
1999 –00	Iberdrola Visiting Professor, University of Valladolid, Spain
2000	Keynote Speaker, IUPAC International Conference on Organometallic Chemistry, Shanghai, China
2002	Gerhard Closs Lecturer, University of Chicago
2003	Faculty Scholar Medal, Pennsylvania State University
2005	Plenary Lecturer, Volkswagen Conference on Nanotechnology in Science, Economy, and Society, Marburg, Germany
2005	Coochbehar Professorship, Indian Association for the Cultivation of Science, Kolkata, India
2005	Plenary Lecturer, IUPAC Workshop on Advanced Materials, Stellenbosch, South Africa
2005	Elected Fellow, American Association for the Advancement of Science
2008	Plenary Lecturer, International Workshop on Defining Issues in Biofuels R&D, Italy
2009	Invited Distinguished Scientist, National Institute for Materials Science, Japan
2010	Adjunct Professor, International Centre for Materials Science, Jawaharlal Nehru Centre for Advanced Scientific Research, India
2010	Plenary Lecturer, Sitges Conference on Statistical Mechanics, Spain
2011	Medal, Chemical Research Society of India (CRSI)

STUDENT HONORS

1997	Student Award, International Precious Metal Institute (to Smita Kacker)
2000	Student Award, International Precious Metal Institute (to April Hennis)
2002	Student Award, International Precious Metal Institute (to Joseph Remias)
2005	Young Investigator Award, American Chemical Society: Division of Inorganic Chemistry (to Walter Paxton)
2006	Rustum and Della Roy Innovation In Materials Research Award (to Walter Paxton)

RESEARCH INTERESTS

Synthetic and mechanistic organotransition metal chemistry; homogeneous and heterogeneous catalysis; environmental chemistry; polymer chemistry; nanotechnology.

Number of Publications: 287; Number of Patents: 23

Electronic copies available from: http://www.researcherid.com/rid/A-9406-2009

ABSTRACT

Surface-centered microbial infestations have been implicated in nosocomial infections, food spoilage, spread of foodborne diseases and biofouling of materials. Hence there is a growing commercial demand for biocidal plastic materials capable of killing microbes found on specialized and daily use surfaces e.g. medical devices & implants, wound dressings, hospital surfaces, food handling surfaces, etc. We will describe new approaches to the design of polymer-based materials that are inherently antimicrobial. These can employed to form long-lasting coatings on reactive and nonreactive surfaces, as well as hydrogels for controlled release of antimicrobial agents.

Slide 1

**Designing Antimicrobial Surfaces,
Polymers and Hydrogels**

Ayusman Sen

**Department of Chemistry
The Pennsylvania State University
University Park, PA 16802 (USA)
E-mail: asen@psu.edu**

J. Am. Chem. Soc., **2006**
Angew. Chem. Int. Ed., **2008**
Langmuir, **2008**

Slide 2

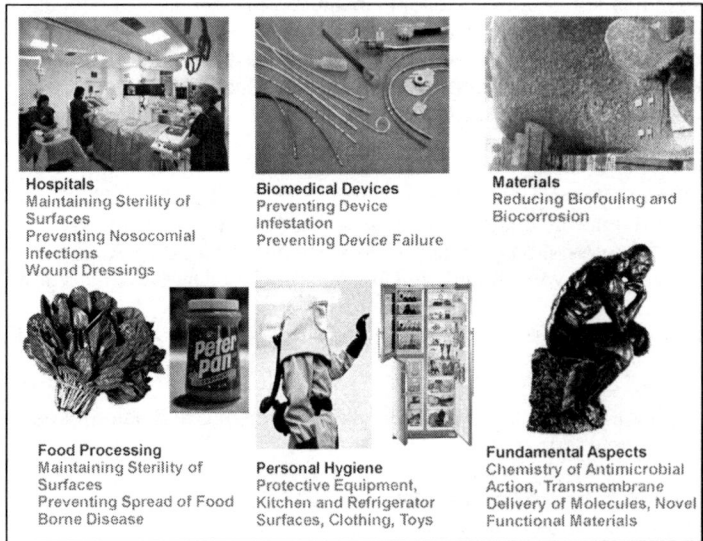

Slide 3

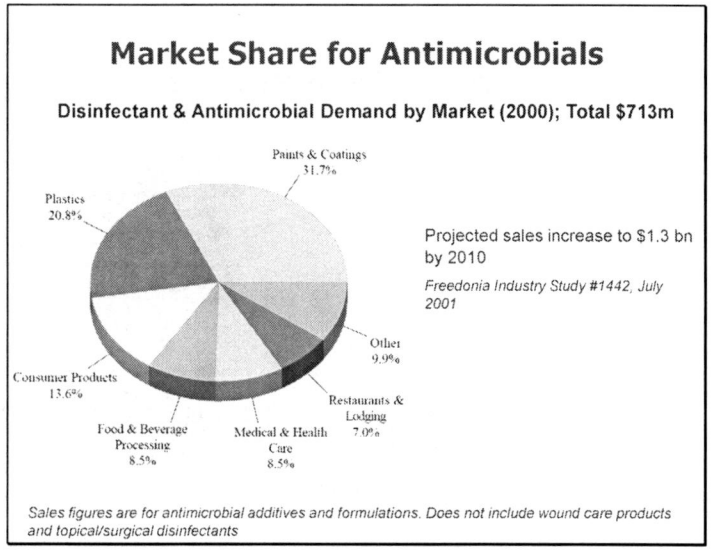

Slide 4

Antimicrobials in Plastic Industry

- Global Market demand for specialty antimicrobials in plastics was $231m in 2001 ($73m in US) with a 5% predicted growth
- Areas of use: 30% in kitchen and bathroom, 25% in food protection (films, trays, house wares), 20% in appliances, 14% in construction, and 12% miscellaneous
- Major environmental and long term toxicity issues with arsenic and tin-based antimicrobials
- Demand for non-arsenic based formulations is predicted to rise at 10-20% per year
- Silver-based formulations are ideally placed to replace use of arsenic and tin-based antimicrobial additives

"Antimicrobials in plastics: a global review" Plastics Additives & Compounding, December 2001

Slide 5

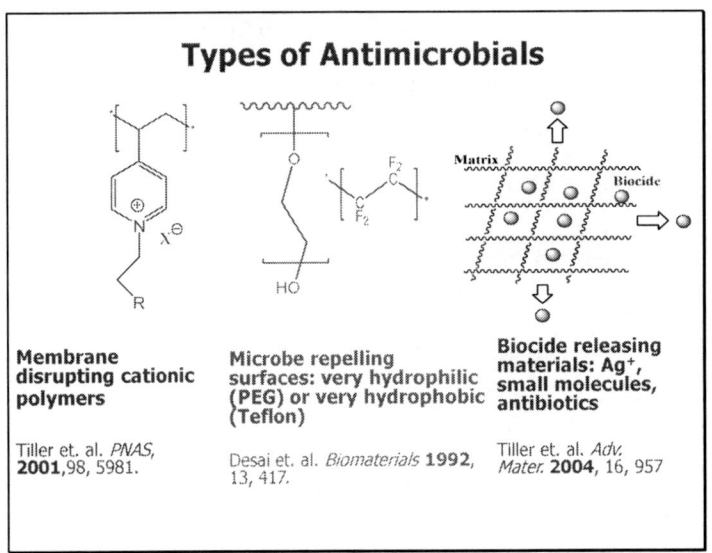

Slide 6

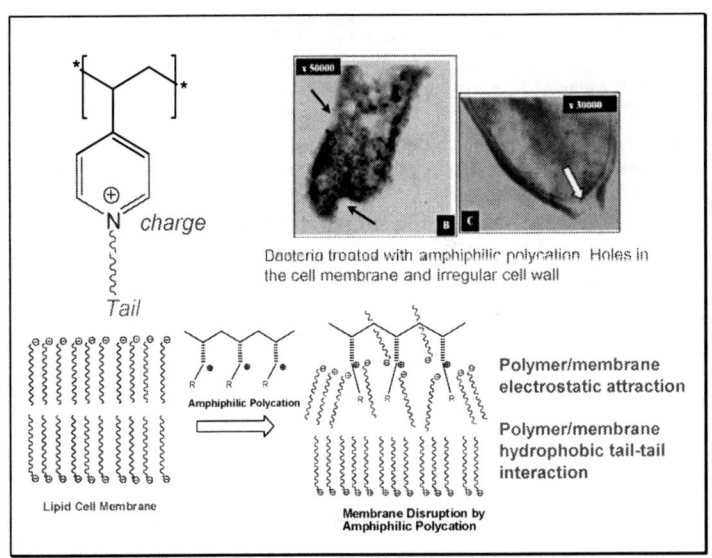

Slide 7

Antibacterial Activity vs. Alkyl Tail Length and % N- alkylation

- None of the polymers are antibacterial at 10% N-alkylation

- Antibacterial activity initially increases with increase in N-alkylation and then plateaus at around 50%. Hence there is no benefit in higher degree of alkylation

- Longer alkyl tails C8 to C16 are inactive towards both E. coli and B. cereus

- Smaller alkyl tails C3 to C6 are active towards E. coli and B. cereus

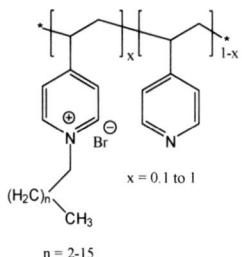

Slide 8

Antibacterial NPVP polymers: *E. coli*

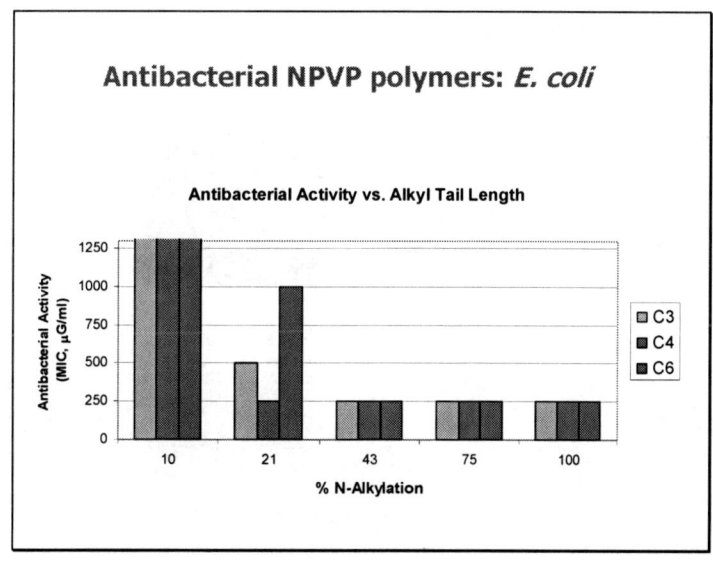

Slide 9

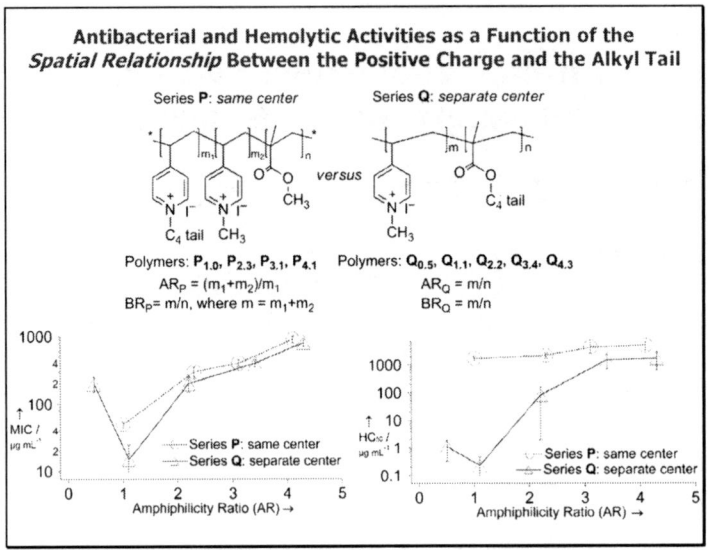

Slide 10

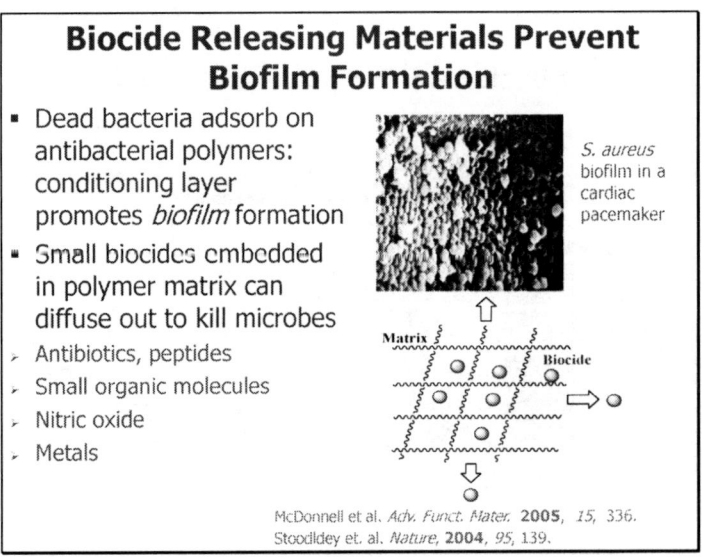

Slide 11

Antimicrobial Silver

- Silver is a well known antimicrobial

- Soluble Ag^+ is believed to be the active species

- Broad antimicrobial activity towards bacteria, virus, fungi

- Effective at low concentration $\sim 10^{-6}$ to 10^{-9} M

- Non-toxic, environmentally friendly

A. D. Russel et. al. *Prog. Med. Chem.* **1994**, *31*, 351

Slide 12

Current Use of Silver Antimicrobials in Consumer Goods

- <u>Apparel & Textiles</u>: work wear, footwear, socks, athletic wear, intimates, general wear
- <u>Consumer Products</u>: cosmetics, air/water filtration, beddings, refrigerators, paper, daily use surfaces, towels, pet care
- <u>Healthcare & Medical</u>: healthcare textiles, diabetic care, wound care, burn care, implants and devices
- <u>Industrial Use & Packaging</u> : wood composites, paper industry, food packaging, consumer goods packaging, paints, antifouling materials in construction, additives for bulk plastics
- <u>Military & Government Services</u>: biological warfare kits, protective garments

Slide 13

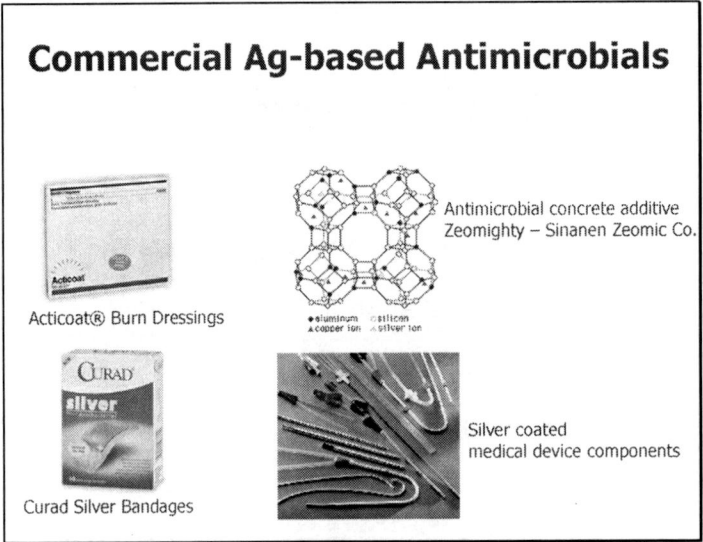

Commercial Ag-based Antimicrobials

Acticoat® Burn Dressings

Curad Silver Bandages

Antimicrobial concrete additive
Zeomighty – Sinanen Zeomic Co.

Silver coated
medical device components

Slide 14

Commercial Antibacterial Paper

- SMEAD: antimicrobial paper products (folders, portfolios). Silver is the active species *http://www.Smead.Com*
- AgION™ antimicrobial office accessories and food touch papers *http://www.agion-tech.com*
- LINTEC corporation (Japan) : antibacterial paper for use in hospital medical records and pharmacy envelopes *http://www.Lintec.Co.Jp/english*
- OG corporation (Japan): additive for antimicrobial paper *http://www.Ogcorp.Co.Jp/en/group/daiwa.Html*

Slide 15

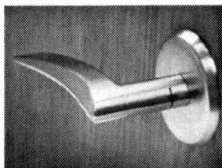

Door handles from Assa Abloy (Sargent) with Agion's silver-based antimicrobial

Staplers from Stanley Bostitch protected with antimicrobials from Agion

Serving utensils from Vollrath (Photo: Agion)

Slide 16

Silver Based Antimicrobials as Plastics Additives

- Inorganic antimicrobial systems based on silver are widely used in Europe and Japan
- Main advantages of silver based materials over organic arsenic and tin formulations is low toxicity, environmental compatibility and excellent thermal stability during high temperature processing
- Global use of silver as biocide in polymeric formulation has risen 600% from 2001 to 2006
- In US silver based antimicrobials in plastics have grown from practically nothing to a significant presence in past few years

Company	Silver Based Antimicrobial Additives for Plastics
AgION	AgION™ Silver
Ciba	HyGate™ HyGentic™
Milliken	AlphaSan® silver
Clariant	Sanitized® Silver
Noble Biomaterials	X-static ®
Argentum Medical	Silverlon™
Medline	SilvaSorb™

"Antimicrobials in plastics: a global review" Plastics Additives & Compounding, December 2002.

Slide 17

Silver Nanoparticles Introduced into Fashion Wear!

Fashion designers and fibre scientists at Cornell University have
designed a silver-impregnated cotton-based garment that can
prevent colds and flu and *never needs washing*!!

Slide 18

Synthesis of Ag-based Antimicrobial Materials

- Blending preformed Ag particles with polymer
- Chemical reduction $Ag^+ \rightarrow Ag^\circ$ *in situ* polymer
- Blending soluble Ag^+-complex molecules with polymer
- Sputtering, embedding, plasma deposition of Ag° nanoparticles onto polymers and other surfaces

- ❖ Complex multi-step synthesis
- ❖ Silver species either poorly soluble or too highly soluble

Slide 19

Goal: Design of Multicomponent Antimicrobial Material Incorporating Silver Halides

- Silver halides offer several advantages
- Intermediate solubility: in between highly soluble silver ions and poorly soluble silver metal
- Very simple synthesis, readily available non-toxic materials
- Simple & novel technique –"on-site precipitation"

❖ Component 1: Membrane disrupting cationic polyvinylpyridinium based polymeric matrix (NPVP)

❖ Component 2: Simple silver salt (AgBr/AgCl)

Slide 20

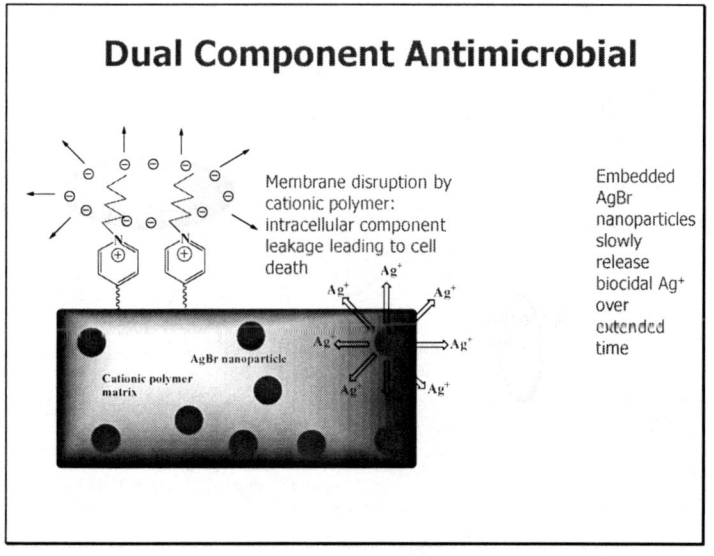

Dual Component Antimicrobial

Membrane disruption by cationic polymer: intracellular component leakage leading to cell death

Embedded AgBr nanoparticles slowly release biocidal Ag+ over extended time

AgBr nanoparticle

Cationic polymer matrix

Slide 21

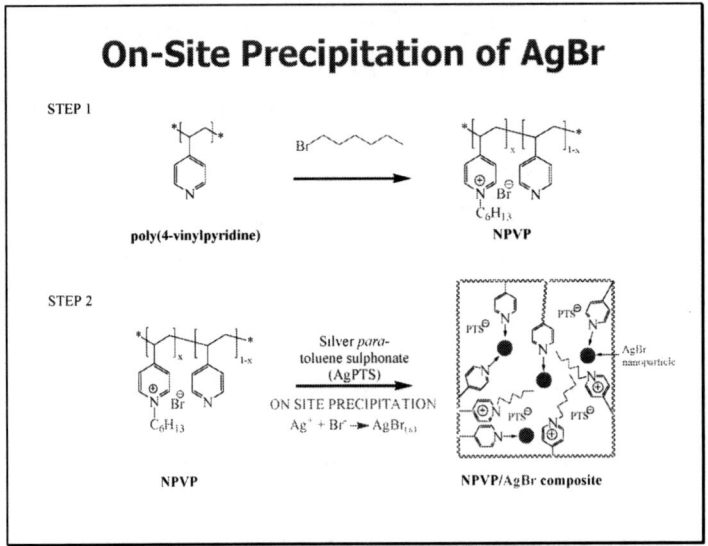

Slide 22

Tailoring AgBr Nanoparticle Size

	% N-alkylation	
	43%	21%
Ag:Br ratio in composite	Average AgBr size (nm)	Average AgBr size (nm)
1:2	10	9
1:1	71	17

- Nanoparticle becomes smaller as the Ag:Br ratio decreases

- Nanoparticle becomes smaller as the degree of alkylation decreases

- Nanoparticle size decreases with the increase in coordinating pyridine / Ag ratio

Slide 23

Antibacterial Activity Assays

- Activity towards gram positive *B. cereus* and gram negative *E. coli*
- Three kinds of assays
  o Surface borne bacteria
  o Air borne bacteria
  o Water borne bacteria

Slide 24

Surface Borne Bacteria: *E. coli*

- Bacteria spread on nutrient agar plates
- 43% NPVP/AgBr nanocomposite impregnated paper placed on plate
- Zone of inhibition measured after 18 h incubation

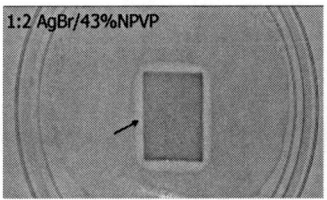
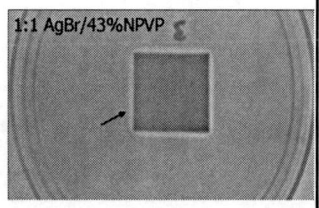

Slide 25

Zone of Inhibition Varies with Particle Size

Ag:Br ratio in composite	AgBr size (nm)	Zone of Inhibition *E. coli* (mm)	Zone of Inhibition *B. cereus* (mm)
1:2	15	3	2
1:1	72	2	1

- Zone of inhibition increase with decrease in particle size
- Smaller particles: higher surface area and hence higher dissolution rate

- **Controlled delivery of biocidal Ag$^+$**

Slide 26

Airborne Bacteria

- *E. coli* mist sprayed on 1:1 AgBr/21% NPVP composite coated surface
- Surface incubated for 18 h to allow bacterial growth
- No growth on/near coated surface

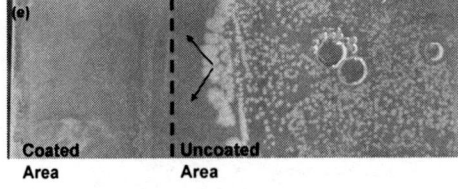

Slide 27

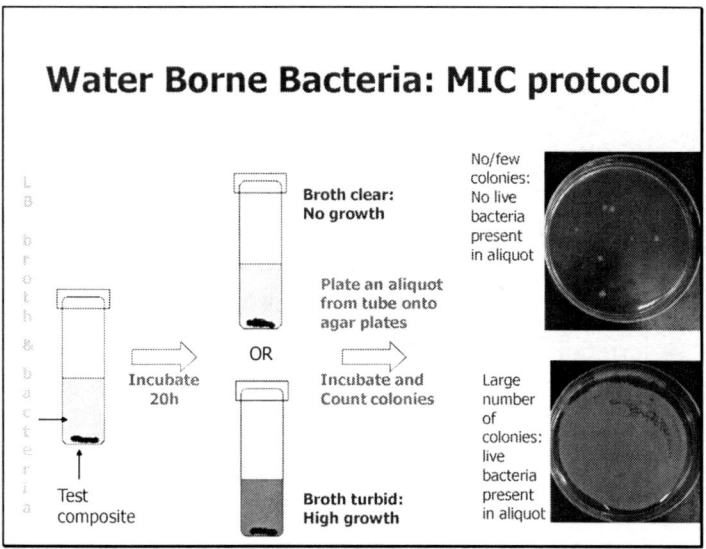

Slide 28

Water Borne Bacteria: MIC Values

Sample	MIC (µg/ml)	
	E. coli	*B. cereus*
1:2 AgBr/21% NPVP	50	50
1:1 AgBr/21% NPVP	50	50
1:2 AgBr/43% NPVP	50	50
1:1 AgBr/43% NPVP	50	50
21% NPVP	1000	1000
43% NPVP	250	250
AgBr	100	100
Na PTS	10,000	10,000
PVP	>10,000	>10,000

- MIC is the lowest concentration (µg/mL) at which a compound will kill > 99% of the added bacteria
- A lower MIC corresponds to a higher antibacterial effectiveness
- AgBr composites had lower MIC than both polymers alone and AgBr alone indicating dual mechanism of action

Slide 29

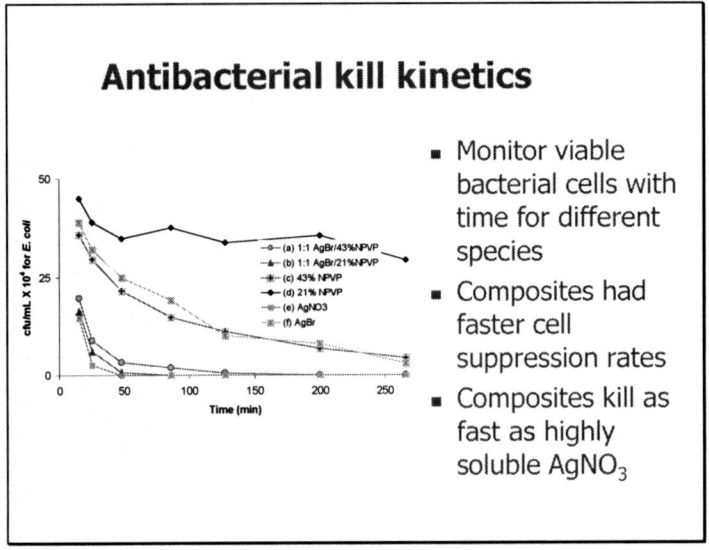

Antibacterial kill kinetics

- Monitor viable bacterial cells with time for different species
- Composites had faster cell suppression rates
- Composites kill as fast as highly soluble AgNO$_3$

Slide 30

Long Lasting Antibacterial Activity

Sample	Composite in 4ml LB broth (µg/ml)	AgBr in 4 ml LB broth (µg/ml)	*E. coli* growth			*B. cereus* growth		
			Day 1	Day 3	Day 10	Day 1	Day 3	Day 10
1:1 AgBr/21%NPVP	500	90	-	-	-	-	-	-
1:2 AgBr/21%NPVP	500	60	-	-	-	-	-	-
AgBr	250	250	-	+	++	-	+	++
21%NPVP	1000	0	+	+++	+++	+	+++	+++

- = no growth, + = small growth, +++ = large growth

Composites have extended time-release properties while retaining potent and fast bacterial suppression

Slide 31

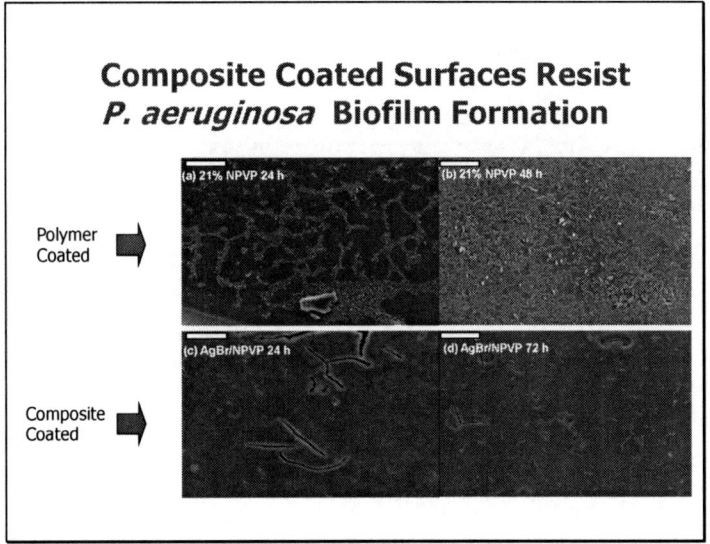

Slide 32

Activity in Mammalian Fluids Towards Methicillin Resistant *Staphylococcus aureus* (MRSA)

Sample	Human Serum	Human Saliva	Human Blood
AgBr/**NPVP** composites	Bactericidal at 150 µg/ml Bacteriostatic at 100 µg/ml	Bactericidal at 100 µg/ml Bacteriostatic at 50 µg/ml	Bactericidal at 200 µg/ml Bacteriostatic at 100 µg/ml
43% **NPVP**	Ineffective	Bactericidal at 1000 µg/ml	Ineffective
21% **NPVP**	Ineffective	Bactericidal at 5000µg/ml	Ineffective

Slide 33

Adhesion of Polymer Coating

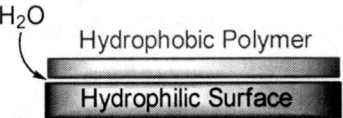

H_2O

Hydrophobic Polymer

Hydrophilic Surface

- Polymer adhesion is one of the primary problems in creating durable coatings
- Incompatibility between hydrophilic surface and hydrophobic polymer leads to poor adhesion
- Water seepage lead to eventual loss of coating material

Methods to improve coating adhesion:

Multipoint Covalent Anchoring of Polymer to Surface

Slide 34

Methoxysilane Based Polymer Coatings

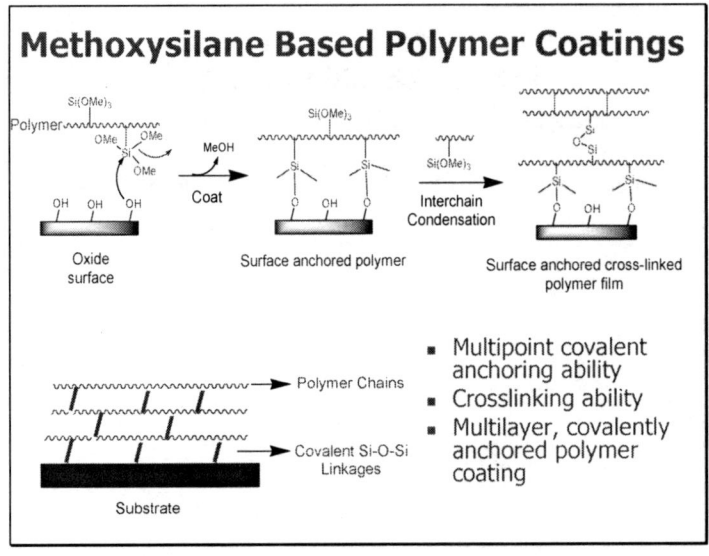

- Multipoint covalent anchoring ability
- Crosslinking ability
- Multilayer, covalently anchored polymer coating

Slide 35

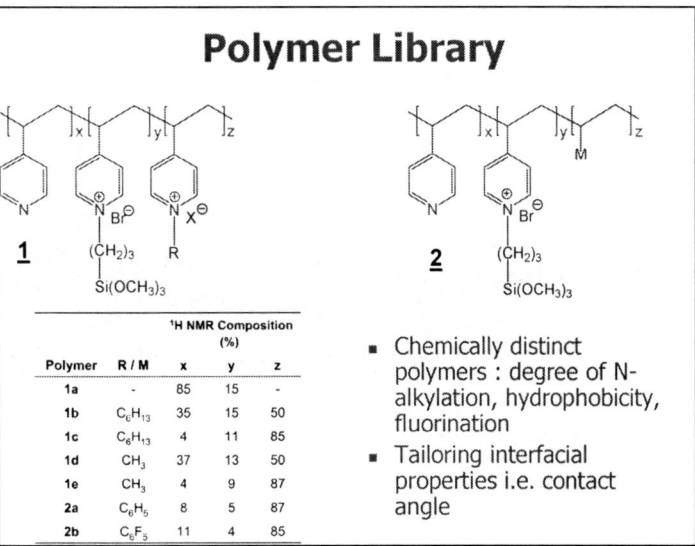

Slide 36

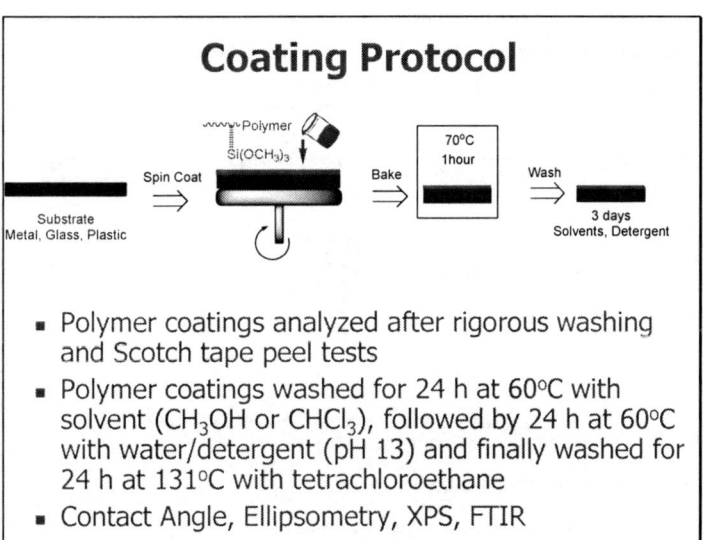

Slide 37

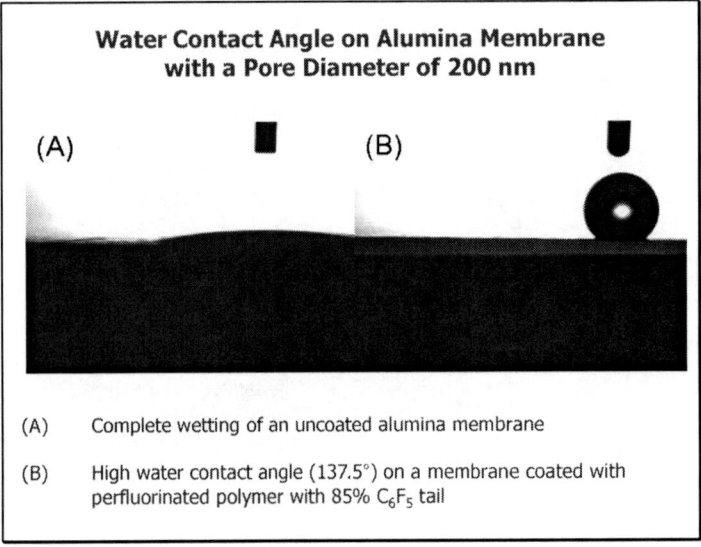

Slide 38

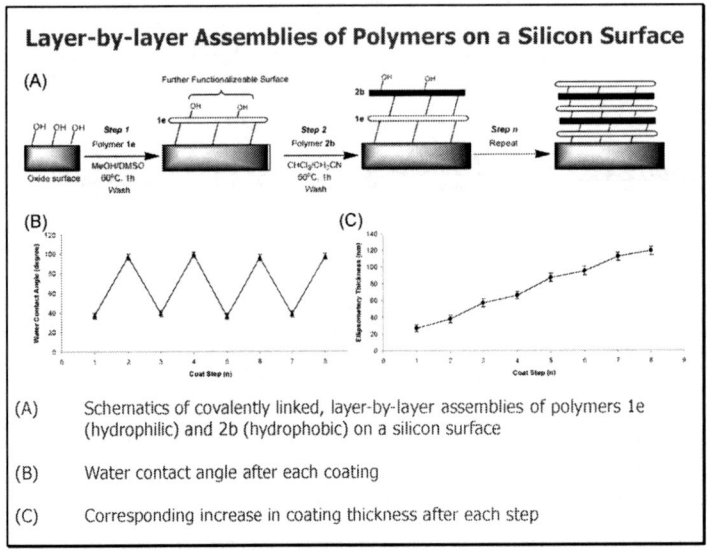

Slide 39

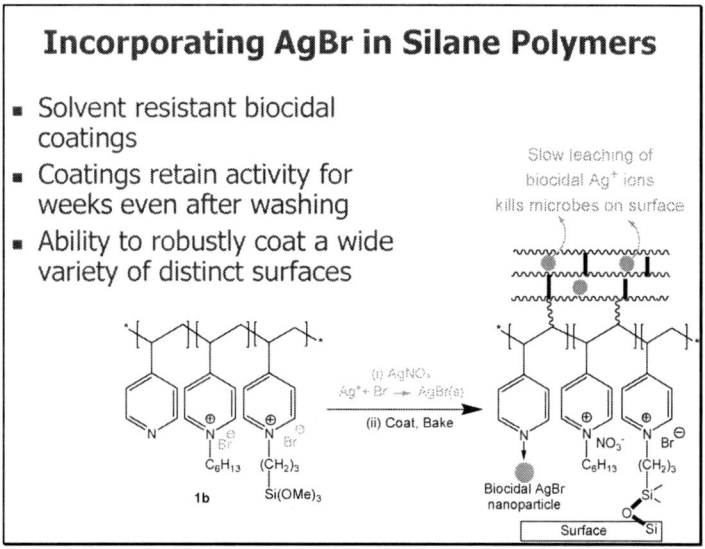

Slide 40

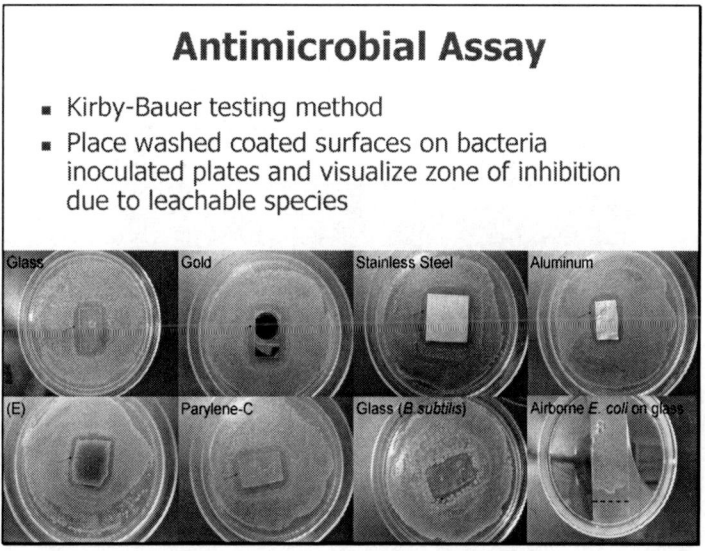

Slide 41

Novel Antimicrobial Hydrogels

Slide 42

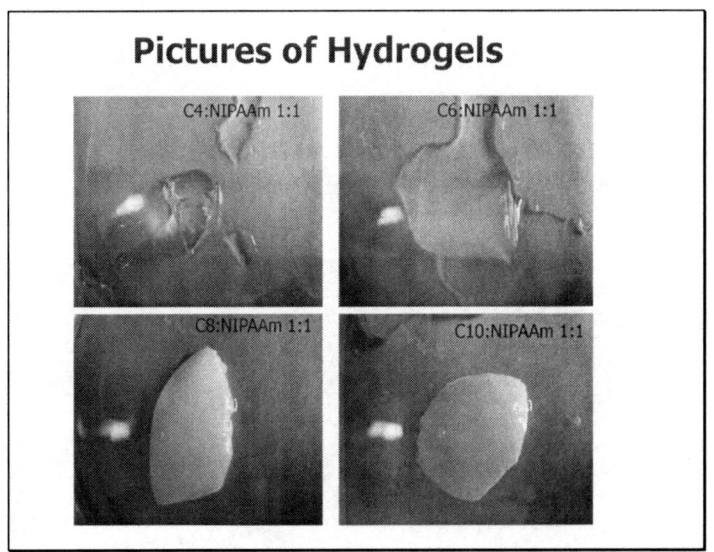

Pictures of Hydrogels

C4:NIPAAm 1:1

C6:NIPAAm 1:1

C8:NIPAAm 1:1

C10:NIPAAm 1:1

Slide 43

Long Term Antimicrobial Test

E: Material effective, no bacterial growth
N: Material no longer effective against bacteria

	E. Coli				B. Cereus			
	Day 1	Day 2	Day 3	Day 15	Day 1	Day 2	Day 3	Day 15
C6:NIPAAm 1:1	E	E	N	N	E	E	N	N
C6:NIPAAm 1:2	E	E	N	N	E	N	N	N
C6:NIPAAm 1:4	N	N	N	N	N	N	N	N
C6:NIPAAm 1:1-AgCl	E	E	E	E	E	E	E	E
C6:NIPAAm 1:2-AgCl	E	E	E	E	E	E	E	E
C6:NIPAAm 1:4-AgCl	E	E	E	E	E	E	E	E

Slide 44

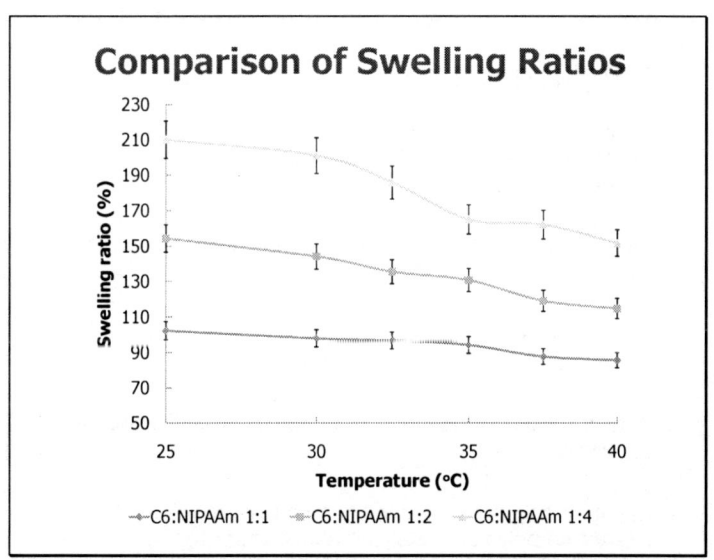

Slide 45

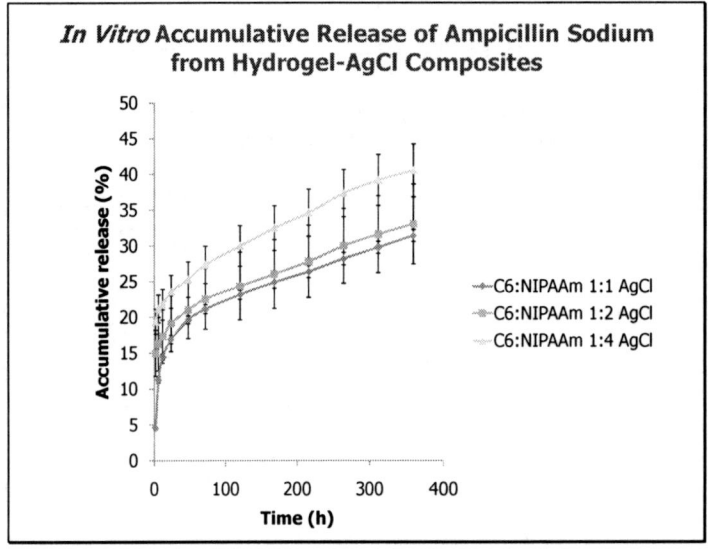

Slide 46

Conclusion

- Novel antibacterial polymeric composites were prepared by simple methods requiring minimal chemical and physical workup

- These form strong, persistent, coatings on a variety of surfaces

- Composites kill both gram positive and gram negative bacteria on surfaces and in solution

- Rate of release of Ag⁺ ion can be tuned by controlling AgX particle size

- On-site precipitation method can be used to make other interesting polymer/metal salt nanocomposites

Slide 47

Formulations

Formulation	Application	Property
Solid AgX/cationic polymer nanocomposites	Blending with bulk polymer e.g. PVC, PE, PMMA, PS	Nanodispersed AgX, excellent blending properties with plastics due to polymer component of nanocomposite
Ethanol based AgX/methoxysilane polymer solutions	Coating textiles (cotton, polyesters, lycra, etc), glass, ceramics, metals, surgical grade plastics, wood, paper	Excellent antimicrobial activity, covalent and non-covalent substrate anchoring, ability to coat any kind of substrate, easy application using widely available technology i.e. spray coating, dip coating
Ethanol based AgX/fluorinated silane polymer solutions	Coating textiles (cotton, polyesters, lycra, etc), glass, ceramics, metals, surgical grade plastics, wood, paper	Excellent water repellency with simultaneous antimicrobial activity, covalent and non-covalent substrate anchoring, ability to coat any kind of substrate

Slide 48

Our Product is Ideally Positioned to Impact Several Areas

Area	Products	Advantages
Plastics	Solid AgX based antimicrobial additives for polymers, wood composites Liquid formulations for coatings	Easy dispersion of our AgBr composite in polymers, methoxysilane polymers form coating on generalized surfaces, ability to tailor polymer structure to enhance bulk property
Wound care	Antimicrobial hydrogels AgX/cellulose dressings Protein/cell repellant fluorinated dressings	Renewable wound dressings, excellent biocompatibility, dual action- prevention of cell adhesion and biocidal effect, formulation binds to cellulose for durability
Diabetes care	AgX containing antimicrobial socks and footwear	Consumer and end user can coat preformed textiles by simple dip/spray coating, AgBr covalently anchored to textile, water repellant coatings possible
Nosocomial Infection	AgX containing hospital textiles e.g. sheets, robes, towels, etc AgX containing surfaces	Proven activity against *MRSA, Pseudomonas, E. coli*
Apparel	All kinds of antimicrobial textiles Water repellant textiles	Ability to coat preformed textiles, simultaneous biocidal and water repellant activity
Surfaces	Coating formulations for generalized surfaces in hospitals, homes, industrial and institutional settings	Ability to coat wide variety of surfaces

Slide 49

NEW ACTIVE INGREDIENT STRATEGIES TO CONTROL FUNGAL GROWTH ON SURFACES

Alex Cornish

Syngenta Crop Protection AG

Schwarzwaldallee 215, Basel CH 4002, Switzerland

Tel: +41 323 9078 Fax: +44 (0)1939 252416 email: alexander.cornish@syngenta.com

BIOGRAPHICAL NOTE

Dr Alex Cornish joined Syngenta in 2005 after working for over 10 years in the Material Protection sector with Zeneca, Avecia and Arch Chemicals. He obtained a PhD at University of Cambridge on the mode of action and biosynthesis of quinoxaline antibiotics in 1982. He is currently Technical Manager for Landscape and Material Protection at Syngenta in Basel.

ABSTRACT

Since 2005, Syngenta has been evaluating its portfolio of agrochemical fungicides in order to explore additional opportunities in Material Protection. Syngenta is the market leader in fungicides for agriculture and continues to invest in existing products by maintaining registrations (for example under the EU Biocidal Products Directive). The company also has an active discovery programme to bring new active substances to market in the agrochemical sector. Re-evaluation of the fungicide portfolio for surface mould control has confirmed that actives have additional utility beyond agriculture as single actives but most often in combination with other active ingredients.

The areas of focus have been control of surface mould growth on wallboard paper, exterior paints and plastics and a summary of recent findings will be reported here.

INTRODUCTION

The Specialty Chemical Sector has ceased to discover and bring to market new fungicides for Material Protection as the market is not large enough to support bespoke discovery and development programmes. However, the agrochemical sector can still support innovation in new active ingredients. The various discovery programmes are targeted at controlling and preventing particular crop diseases, but new actives may nevertheless have utility in the Material Protection Sector. Indeed, Chlorothalonil, folpet and carbendazim are all fungicides that were orginally developed for agriculture, but have made important contributions in Material Protection as fungicides for paint and plastic.

Chlorothalonil and folpet are examples of broad-spectrum, multi-site compounds that serve to inhibit multiple enzyme systems in the fungi. Such broad spectrum activity is usually exactly what is needed for a Material Protection Product that must be able to control numerous moulds that cause surface defacement and spoilage. However, the current trend is to develop new molecules that are highly selective for a particular enzyme and target a particular sub-sets of plant pathogens that have no relevance to Material Protection, so development of safe, multi-site compounds for the Material Protection Sector becomes even less likely.

The objective of the present work was to investigate further agrochemical fungicides in the context of Material Protection targets to discover they had potential utility for surface mould control.

CLASSES OF COMPOUNDS AND MODES OF ACTION

The following fungicides in Table 1 were considered in the present work

Table 1. Fungicides considered in the present study					
	Propiconazole	Defenoconazole	Thiabendazole	Azoxystrobin	Fludioxonil
Abbreviation used in text	PPZ	DFZ	TBZ	AZO	FDL
Chemical Class	Triazole	Triazole	Benzimadazole	Strobilurin	Phenyl pyrrole
Mode of action	Block ergosterol biosynthesis	Block ergosterol biosynthesis	Disrupts tubulin assembly	Respiration inhibitor	Disrupts osmoregulation
Existing Material Protection end uses	Wood decay Wood stains Sapstain	Not used	Paper and more recently adhesives and paint in teh EU	Wallboard	Wallboard
Physical from	Viscous liquid	Solid	Solid	Solid	Solid

The compounds selected have very specific modes of action and all possess extensive toxicology and ecotoxicology packages that are mandatory for product used on food crops.

STRATEGIES FOR COMBINATION DEVELOPMENT

Three target areas for surface mould control were selected for investigation. These were paper, paint and plastic (plasticised PVC). In order to identify the best combinations, the active ingredients were grouped in various combinations and ratios (usually 80:20, 50:50 and 20:80 for a given pair and incorporated into the test matrix pilot scale equipment for paper and lab scale for paint and plastic. The test pieces were then evaluated for resistance to surface growth of fungi using recognised test methods (see below).

NEW COMBINATIONS FOR MOULD CONTROL IN PAPER

Thiabendazole, azoxystrobin and fludioxonil combinations were tested in paper using the ASTM G21 test method that was designed for plastic but was selected for paper because it provides a stringent fungal challenge. Neither propiconazole nor difenaconazole were evaluated in this substrate (Wolf et al, 2008 Gypsum World paper 1).

The screen (Figure 1) showed that paper treated with the AZO/TBZ combination gave superior mould control compared with commercial mould resistant wallboard paper, but addition of a small amount of FDL gave much better mould control to the point where no growth was observed. This combination of three actives also controlled *Stachobotrys chartarum* and was subsequently commercialised as Sporgard WB (Woods et al 2009).

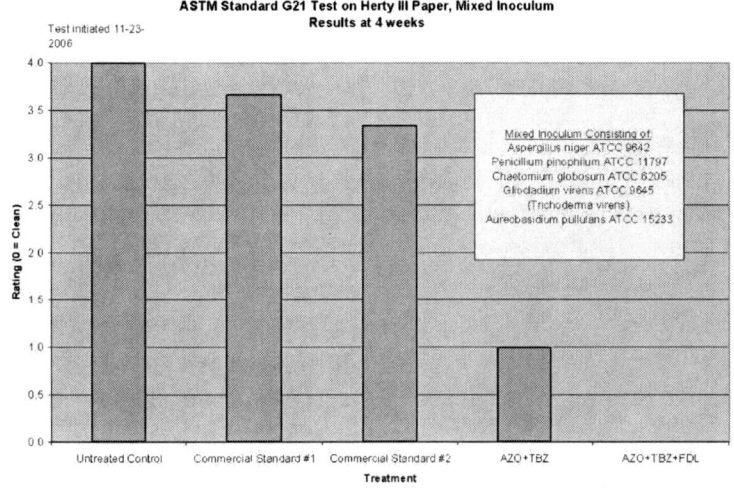

Figure 1. Mould-resistant properties of fungicide-treated paper.

Commercial samples of mould-resistant wallboard and papers treated with AZO/TBZ and AZO/TBZ?FDL combinations were evaluated for mould resistance using the ASTM G21 test (Wolf et al 2008)

PAINT

The selected active ingredients were tested alone and in various combinations in an acrylic flat paint. Dosed paint samples were painted (2 coats) onto unprimed Southern Yellow Pine panels and exposed for 36 months (North facing) in Clearwater, Florida. Paints were dosed with fungicides at a total active ingredient loading of 0.25 or 0.5% by weight. The algicide diuron was also added to all fungicides samples at 0.25%.

The results showed that on a 0-10 rating scale the single actives in general did not score above 6 compared to fungicide-free controls (scores 2-4), but most of the combinations gave an acceptable 8 or higher. There was clear evidence that actives such as TBZ and FDL can give a booster effect when mixed with other actives that do not deliver acceptable performance as solo compounds.

PLASTIC

As an initial screen, the various lead fungicides were incorporated into plasticised PVC as dispersions in di-*iso*-nonyl phthalate anad evaluated using the ASTMG21 test. TBZ emerged as the lead candidate, but caused "blooming" due to migration of the active ingredient to the surface at concentrations above 0.15 by weight.

However, by adding FDL it proved possible to reduce the TBZ concentrations significantly below 0.1% whilst retaining antifungal activity (Table 2).

Table 2.

Initial evaluation of TBZ and FDL as fungicides for plasticised PVC.

Full details are given in Cornish & Herbst 2007

	Fungicide content (% by wt)		Resistance to fungal growth
Sample	FDL	TBZ	sing ISO 846 test
Control	0	0	100%
1	0	0.1	100%
2	0.02	0.08	40%
3	0.025	0.075	0%
4	0.033	0.067	0%
5	0.05	0.05	0%

CONCLUSIONS

During the course of this work, we have discovered that a number if agrochemical fungicides have utility as material protection agent. They differ from classic Material Protection fungicides in that they are not usually effective as stand alone biocides. However, they can deliver powerful effects when used judiciously in combination with other active ingredients.

The additional utility of these compounds is worth noting as the number of available fungicidal active ingredients continues to decline, particularly as they are already supported by extensive toxicology and ecotoxicology packages

REFERENCES

1. Wolf, H. Woods, T. Valentine S., Vink R. & Cornish A. Development of technologies for mould-resistant wallboard: Part 1 – Overview and test methods. Global Gypsum Magazine, September 2008, pp 22-28.

2. Woods T., Valentine S., Wolf H., Bonnett P., Cornish A. & Vink R. Optimised mould control in wallboard- Part II: Sporgard WB.Global Gypsum Magazine March 2009 pp 20-24.

3. Cornish A & Herbst H. International Patent Application (PCT) - PCT/GB2007/004836 .
    Relating to: Formulations which comprise a plasticizer and articles made therefrom and/or coated therewith wherein the formulation comprises a plasticizer and a fungicidally effective amount of a fungicide

ACKNOWLEDGEMENTS

I would like to acknowledge the contribution of my Syngenta colleagues Dr Hanno Wolf, Dr Thomas Wood , Dr Anja Greiner and Robert Vink. I would also like to thank Dr Heinz Herbst for his collaboration on PVC.

RELATING FUNCTION TO CLAIMS FOR ANTIMICROBIAL PROPERTIES IN TEXTILE APPLICATIONS

Pete Askew

IMSL

Pale Lane, Hartley Wintney, Hampshire, RG27 8DH, UK

Tel: 01252 62 76 76 Fax: 01252 62 76 78 email: peter.askew@imsl-uk.com

BIOGRAPHICAL NOTE

Pete Askew has been working as a microbiologist since 1977. For the first 2 years he worked in the food industry before joining the microbial ecotoxicology unit of ICI's agrochemicals business. In 1988 he became company microbiologist for ICI Paints and in 1996 left to form IMSL. IMSL is a specialised microbiological testing and consultancy service based in laboratories and offices near Fleet UK. The company has a branch office in Potsdam, Germany. Work is confined to industrial systems (eg coatings, adhesives, plastics, textiles, spin finishes, paper and disinfectants). No standard food and water testing is undertaken. IMSL is highly active in the development of testing methodologies for the determination of the performance of treated articles against a range of microorganisms including bacteria and viruses.

Pete is a member of all of the major microbiological societies and is Secretary General of the International Biodeterioration Research Group and is the chaiman of the Textiles and Plastics work groups. As well as qualifications in microbiology, he is an Associate of the Oil and Colour Chemists Association and is the consultant to the OECD on treated articles.

ABSTRACT

Biocides have been used in the production and for the protection of finished textile-based goods for many years. However, the last few years has seen the sustained growth in the number of textiles that are treated to provide antimicrobial and hygienic effects reaching the market. In response, authorities worldwide are putting into place structures to regulate the claims made for such products. While there are a number of methods available for measuring antibacterial properties associated with treated textiles, in many cases these only provide information about potential activity. The exposure scenarios provided by these methods presents the microorganisms in aqueous suspensions with contact times of 24 hours and exposure temperatures of 35 - 37°C. Rarely do these methods reflect well the conditions under which microorganisms will come into contact with them in actual use.

This paper examines a systematic approach to producing data to support claims made for antimicrobial treated textiles. A series of test protocols are described which can produce data linked to specific exposure scenarios and help to provide evidence to support claims.

Introduction

Textiles have been produced and used by humans for millennia but until relatively recently their composition has been dominated by natural materials, usually from fibres derived from either plants or animals. However, in the last 100 years, natural fibres have been mixed with man-made materials created either from polyolefins such as Nylon or modified natural materials such as Viscose and Rayon. In many cases, man-made fibres have either replaced traditional natural fibres in modern textiles or added properties which could not previously be obtained (eg a high degrees of elasticity) and other additives are being used increasingly to add new functionality (eg hydrophobicity and stay-fresh properties).

Despite the emergence of many man-made materials, the biodeterioration of textiles remains a problem, with microorganisms causing discolouration of / odour in finished goods and loss of functionality (eg loss of tensile strength in canvas - Ref 1). Similarly, microbial action in textile manufacturing processes causes losses in productivity as well as function (eg blockage of applicators of spin finishes / weaving ancillaries by microbial growth / detached microbial biofilms resulting in either yarn being produced without antistatic agents / lubricants or areas of localised damage due to overheating on the loom).

The biodeterioration of finished textiles is usually associated with fungi although a wide range of microorganisms cause problems in the manufacturing environment. Growth occurs on the finished goods when the material is exposed to either high humidity or free water during use and storage (Ref 2). The resultant growth causes either marking and discolouration (Plate 1) or, eventually, physical damage. Such discolouration and spoilage might range from small blemishes to the severe staining / musty odours associated with mildew on tents that have been stored without being dried properly first, possibly even resulting in structural failure. Growth of microorganisms is usually prevented by ensuring that insufficient moisture is present but, where exposure to moisture is expected during service or where less than ideal storage conditions prevail, goods can be treated to prevent growth and spoilage (Ref 2). This use of preservative technology is well understood and a number of standard test protocols exist that can be used optimise performance and predict durability (see Tables 1). However, in recent years a number of textile-based goods have been produced which include antimicrobial properties which are not intended merely to prevent deterioration in service, but which are intended to provide an antimicrobial function in use. These articles include items of clothing fortified with microbicides either to prevent odours being formed from human sweat or to prevent cross-infection in clinical environments. Test protocols for these are less well defined and often fail to predict their performance in service.

Table 1: Methods used to Examine the Resistance of Textiles to Biodeterioration

Reference	Title	Description	Major Principle
EN 14119	Testing of textiles -Evaluation of the action of microfungi	The test is designed to determine the susceptibility of textiles to fungal growth. Assessment is by visual rating and measurement of tensile strength.	Agar plate test
AATCC 30	Antifungal activity, Assessment on textile materials: mildew and rot resistance of textile materials	The two purposes of the test are to determine the susceptibility of textiles to microfungi and to evaluate the efficacy of fungicides on textiles	Agar plate test
DIN 53931	Testing of textiles; determination of resistance of textiles to mildew; growth test	The test determines the efficacy of treatments for prevention of fungal growth on / in textiles. It also allows the performance testing of a treatment after UV irradiation , leaching *etc*.	Agar plate test
MIL-STD-810F	Environmental Engineering considerations and laboratory tests; Method 508.5 FUNGUS	The purpose of the method is to assess the extent to which a material will support fungal growth and how performance of that material is affected by such growth.	Humid chamber test (90 to 99% humidity)
BS 6085 :1992	Determination of the resistance of textiles to microbial deterioration	The purpose of the method is to assess the extent to which a material will support fungal / bacterial growth and how performance of the material is affected by such growth by visual assessment and measurement of tensile strength	a) soil burial test; b) agar plate test, c) humid chamber test
EN ISO 11721-1	Textiles - Determination of resistance of cellulose-containing textiles to micro-organisms - Soil burial test- Part 1: Assessment of rot retarding finishing	The test is designed to determine the susceptibility of cellulose containing textiles against deterioration by soil micro-organisms by visual assessment and measurement of tensile strength	Soil burial test
EN ISO 11721-2	Textiles - Determination of resistance of cellulose containing textiles to micro-organisms - Soil burial test- Part 2: Identification of long-term resistance of a rot retardant finish	The test identifies the long-term resistance of a rot-retardant finish against the attack of soil inhabiting micro-organisms. It allows the distinction between regular long-term resistance and increased long-term resistance to be made	Soil burial test
BS 2011 : Part 2.1J (IEC 68-2-10)	Basic environmental testing procedures	Mould growth test to show the susceptibility of a material towards colonisation by fungi.	Humid chamber test (90 to 99% humidity)
AS 1157.2 - 1999	Australian Standard - Methods of Testing Materials for Resistance to Fungal Growth Part 2: Resistance of Textiles to Fungal Growth. Section 1- Resistance to Surface Mould Growth.	Test specimens are inoculated with a suspension of spores of *Aspergillus niger* and then incubated on the surface of a mineral salts based agar for 14 days and then assessed for growth. Both leached and unleached specimens are examined. Glass rings are employed to hold the specimens in intimate contact with agar when necessary. Specimens are examined for the presence of surface mould growth.	Agar plate test

Reference	Title	Description	Major Principle
AS 1157.4 - 1999	Australian Standard - Methods of Testing Materials for Resistance to Fungal Growth Part 2: Resistance of Textiles to Fungal Growth. Section 2 - Resistance to Cellulolytic Fungi.	Test specimens are inoculated with a suspension of spores of *Chaetomium globosum* and then incubated on the surface of a mineral salts based agar for 14 days and then assessed for growth. Both leached and unleached specimens are examined and exposed samples are subjected to a tensile strength test. Glass rings are employed to hold the specimens in intimate contact with agar when necessary.	Agar plate test
AS 1157.3 - 1999	Australian Standard - Methods of Testing Materials for Resistance to Fungal Growth Part 2: Resistance of Cordage and Yarns to Fungal Growth.	Test specimens are inoculated with a suspension of spores of *Chaetomium globosum* and then incubated on the surface of a mineral salts based agar for 14 days and then assessed for growth. Both leached and unleached specimens are examined and exposed samples are subjected to a tensile strength test.	Agar plate test (other vessels containing media are employed for large specimens).

Antimicrobial Textiles

The range of applications addressed by textiles designed to exhibit an antimicrobial / hygienic effect is extensive. Products have been developed that are intended to enhance the freshness of garments through the prevention of odour resulting from microbial action on human sweat compounds whether in normal, outer day-wear (shirts *etc*), underwear or sports clothing. Other applications include antimicrobial wound dressings intended either to enhance wound healing or extend the intervals between the changing of dressings. Similarly, medical uniforms have been treated in an attempt to reduce the incidence of hospital acquired infections such as drapes used during surgical procedures and soft furnishings used in the hospital environment (*eg* curtains and seating fabric). Socks containing copper-based additives have been used to treat persistent ulcers in individuals with diabetes and pyjamas that claim to eliminate MRSA from the skin are marketed for use prior to admission to hospital. The range of claims is extensive, however, relatively few standard test protocols exist to help substantiate them and fewer still provide useful models of end use (see Table 3).

Methods used to Measure Activity

It can be seen from Table 2 that there are two major forms of test for the microbiological effects of treated textiles. In the first, typified by the qualitative component of JIS L1902, samples of textile are placed onto agar plates which have been inoculated with bacteria and are then incubated. The intention is that intimate contact between the textile, the bacteria and the growth medium will result in the inhibition of growth either immediately adjacent to the textile or in an area around the textile should any antimicrobial agents that have been employed become dissolved in the growth medium. These methods are generally acknowledged as being non-quantitative although they could potentially be employed as assays of certain antimicrobial products in the same manner that such techniques are used for some antibiotics. This can be useful as a screening tool and for investigating the effect of wash cycles *etc*. Such methods are widely employed in the textile industry as they provide a highly graphic representation of antimicrobial activity. However, this can lead to the misunderstanding of the scale of effect seen (bigger zones of inhibition looking better) and the implications that mobility of active ingredient has on service life (*eg* loss during laundering). For immobilised active ingredients and for active ingredients that interact with proteins such as silver ion donors, these methods often have little utility either because the active ingredient cannot migrate to form a zone or because the high concentration of proteins present in the growth medium deactivates the antimicrobial agent. Although these techniques are considered to be unsuitable for 'quantifying' the effect of the antimicrobial effects of treated textiles there are some disciplines in which they may provide data which is more relevant to the effect claimed than that delivered by a fully quantitative technique. For example, the interaction with a microbial population on a semi-solid, protein rich medium can provide a useful model for wound dressings.

In addition to the qualitative tests, it can be seen from Table 2 that there are at least 4 techniques which provide quantitative data on the effect of treated textiles on bacteria. These are typified by the method described in ISO 20743 in which samples are inoculated with suspensions of bacteria and then incubated for a specified time before being examined for the size of the population present. The methods differ in the form of the suspension medium, number of replicates examined, test species and, to a certain extent, conditions for incubation. The methods described in AATCC-100 and JIS L 1902 appear to be the most commonly employed although the absorption method in ISO 20743 is being cited with more frequency.

It can also be seen from Table 2 that no fully quantitative methods yet exist for the examination of the effect of treated textiles on fungi. All of the protocols described are zone diffusion assays of one form or another but, as with the antibacterial properties, these may be sufficient to substantiate certain claims. A quantitative method is under development in Japan, however, in the form of an early draft ISO standard, which employs the measurement of ATP to predict / quantify the early stages of fungal spore germination and growth.

As discussed above, these methods are employed for a wide range of claims made by a number of antimicrobial treatments and have emerged mainly from the need to determine the performance of textiles associated with healthcare applications (*eg* dressings, uniforms of medical staff *etc*) where relatively large or rapid antibacterial effects are required. However, the published methods are less well suited to exploring effects associated with either small bacterial populations or relatively low levels of metabolic activity, for example the transformation of sweat and sebum into odour compounds (Ref 3). While the formation of body odour is attributed to a number of bacterial species associated with human skin, fewer data are available for such transformations on textiles worn close to the skin. Used sportswear often develops musty odours before being laundered and this is often associated with the presence of pseudomonads on the damp clothing. This odour does not usually develop when worn. The transformation of steroids *etc* to the main components of body odour appears to be attributed to a number of bacterial species, such as coryneform bacteria (Ref 4), probably functioning within consortia on the skin / in sebum glands. Some transformations may also occur on clothing. In contrast, a smaller range of bacteria are associated with foot odour, with a number Gram positive species, including staphylococci, being implicated (Ref 5).

Table 2: Methods used to Examine the Antimicrobial Activity of Textiles

Reference	Title	Description	Major Principle
ASTM E2149-01	Standard Test Method for Determining the Antimicrobial Activity of Immobilized Antimicrobial Agents Under Dynamic Contact Conditions	Dynamic shake flask test. Test material is suspended in a buffer solution containing a known number of cells of *Klebsiella pneumoniae* and agitated Efficacy is determined by comparing the size of the population both before and after a specified contact time.	Relies on either diffusion of antimicrobial mater from treated material into the cell suspension or interaction between the population and the surface of the material in suspension.
AATCC 147-1998	Antibacterial Activity Assessment of Textile Materials: Parallel Streak Method	Agar plates are inoculated with 5 parallel streaks (60 mm long) of either *Staphylococcus aureus* or *K pneumoniae*. A textile sample is then placed over the streaks and in intimate contact with the surface of the agar and incubated. Activity is assessed based on either the mean zone of inhibition over the 5 streaks or the absence of growth behind the test specimen.	Zone diffusion assay.
AATCC 100-1999	Antibacterial Finishes on Textile Materials: Assessment of	Replicate samples of fabric are inoculated with individual bacterial species (*eg Staph aureus* and *K pneumoniae*) suspended in a nutrient medium. The samples are incubated under humid conditions at 37°C for a specified contact time. Activity is assessed by comparing the size of the initial population with that present following incubation. A neutraliser is employed during cell recovery.	Cell suspension intimate contact test.
XP G 39-010	Propriétés des étoffes - Étoffes et surfaces polymériques à propriétés antibactériennes - Caractérisation et mesure de l'activité antibactérienne	Four replicate samples of test material are placed in contact with an agar plate that has been inoculated with a specified volume of a known cell suspension of either *Staph aureus* and *K pneumoniae* using a 200g weight for 1 minute. The samples are then removed. Duplicate samples are analysed for the number of viable bacteria both before and after incubation under humid conditions at 37°C for 24 hours. A neutraliser is employed during cell recovery.	Cell suspension intimate contact test.
JIS L 1902	Testing Method for Antibacterial Activity of Textiles Qualitative Test	Three replicate samples of fabric, yarn or pile / wadding are placed in intimate contact with the surface of agar plates that have been inoculated with a cell suspension of either *Staph aureus* or *K pneumoniae* and incubated at 37°C for 24 - 48 hours. The presence of and size of any zone of inhibition around the samples is then recorded.	Zone diffusion assay.

Reference	Title	Description	Major Principle
JIS L 1902	Testing Method for Antibacterial Activity of Textiles Quantitative Test	Replicate samples of fabric (6 of the control and 3 of the treated) are inoculated with individual bacterial species (*eg Staph aureus* and *K pneumoniae*) suspended in a heavily diluted nutrient medium. The samples are incubated under humid conditions at 37°C for a specified contact time. Activity is assessed by comparing the size of the initial population in the control with that present following incubation. No neutraliser is employed during cell recovery.	Cell suspension intimate contact test.
SN 195920	Examination of the Antibacterial Effect of Impregnated Textiles by the Agar Diffusion Method	Four replicate samples of fabric (25 ± 5 mm) are placed in intimated contact with a solid nutrient medium in a petri dish. The samples are then overlaid with molten solid nutrient media which has been inoculated with a cell suspension of either *Staph aureus* or *E coli*. The plates are then incubated for between 18 and 24 hours and the plates are then assessed as described in prEN ISO 20645 above.	Zone diffusion assay
ISO 20645	Textile Fabrics - Determination of the antibacterial activity - Agar plate test	Four replicate samples of fabric (25 ± 5 mm) are placed in intimated contact with a solid nutrient medium in a petri dish. The samples are then overlaid with molten solid nutrient media which has been inoculated with a cell suspension of either *Staph aureus*, *Escherichia coli* or *K pneumoniae*. The plates are then incubated for between 18 and 24 hours and the plates are then assessed for growth based on either the presence of a zone of inhibition of > 1 mm or the absence / strength of the growth in the media overlaying the test specimen.	Zone diffusion assay
SN195924	Textile Fabrics - Determination of the Antibacterial Activity: Germ Count Method	Fifteen replicate samples (each replicate is comprised of sufficient specimens of 25 ± 5 mm to absorb 1 ml of test inoculum) are inoculated with cells of either *E coli* or *Staph aureus* suspended in a liquid nutrient medium and incubated in sealed bottles for up tp 24 hours at 27°C. After 0, 6 and 24 hours, 5 replicate samples are analysed for the size of the viable population present. A neutraliser is employed. An increase of 2 orders of magnitude of the population exposed to a control sample is required to validate the test. The method defines a textile as antibacterial if no more than a specified minimum level of growth is observed after 24 hours in 4 of the 5 replicate groups of samples.	Cell suspension intimate contact test.
SN195921	Textile Fabrics - Determination of Antimycotic Activity: Agar Diffusion Plate Test		Zone diffusion assay
ISO 20743	Textiles - Determination of antibacterial activity of antibacterial finished products	1- Adsorption Method - see JIS L 1902 - Quantitative Method 2 - Transfer Method - Bacteria are transferred from a cell suspension to a sample of textile using the surface of an agar plate under defined conditions to simulate bacterial transfer from skin. 3 - Printing method - Bacteria are transferred from a cell suspension to a sample of textile via a membrane filter under defined conditions to simulate the transfer of bacteria from dry environmental surfaces *etc*.	

In contrast to a number of antibacterial effects (*eg* the elimination of pathogens carried in splashes of contaminated body fluids on medical uniforms - Ref 6 - see below, prevention of growth of bacteria in dressings on suppurating wounds, *etc*) the control of odour in textiles caused by bacteria requires only a reduction in the rate of their metabolism to reduce the production of odour compounds. Thus, textiles that can inhibit their growth (rather than necessarily killing the populations) can affect the generation of odour within a garment. Ideally, any antimicrobial agent used for such purposes should not migrate to the skin. Textiles that control odour within the textile only, do not at present require registration under the EU Biocidal Products Directive. However, if the antimicrobial agents employed migrate to the skin registration is likely to be required (Ref 7). It is important therefore, that the choice of antimicrobial agent and the concentration of that agent is chosen with care. It is equally important that the method that is employed to measure activity is appropriate to the claim being made. As the control of odour requires relatively subtle antibacterial effects, the use of protocols that apply unrealistic demands on the system under test should be avoided, such that the textile is not 'over treated' and results in migration of antimicrobial agent from the textile to the skin. For example, systems that employ silver ions to create antimicrobial effects are intended for relatively clean systems. The use of test protocols that employ a high concentration of nutrients, which might be suitable for healthcare scenarios, result in the deactivation of the silver ions released before they interact with microbial cells (they act as soiling agents as well as nutrients). For this reason a number of variants of the basic tests described above are under development that model better the end use of a textile and ensure that treatments are appropriate to the effect being targeted.

As discussed, many of the methods described in Table 2 were developed with medical applications in mind. Despite this, they provide a poor model for many end uses as they present a bacterial population as a cell suspension and maintain a wet system during the contact interval. Clearly this is unrealistic for most end uses. The relevance of a test method to the intended application is critical to be of scientific value. For this reason a number of protocols have been developed which simulate certain exposure scenarios. Once such scenario (Ref 6) is to examine the impact of a treatment of a textile on the survival of bacteria carried in small splashes of contaminated liquids when they alight on the surface (such as might occur and go unobserved during patient care in a hospital). The protocol is illustrated in Figure 1 and a model data-set is presented in Figure 2. It can be seen that in this example the treatments have resulted in an increase in the rate of decline of a microbial population that has come into contact with the textile (simulating for example a splash of sputum on a medical uniform) compared with an untreated textile. Such effects could bring a benefit in reducing the risk of transfer of microbial contamination from patient to patient in a clinical setting.

Figure 1: Method to Model Activity Against Bacteria Carried in Splashes of Contaminated Fluids

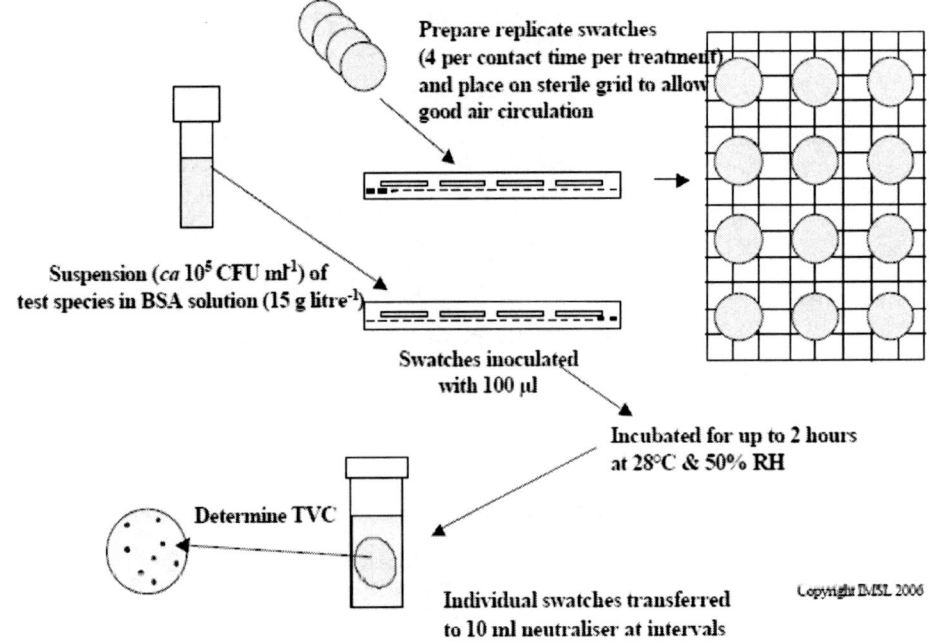

Figure 2: Activity of Treated Textiles Against *Staphylococcus aureus* Carried in a Splash of a Contaminated Simulated Body Fluid

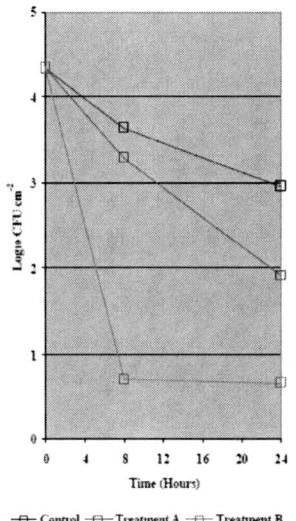 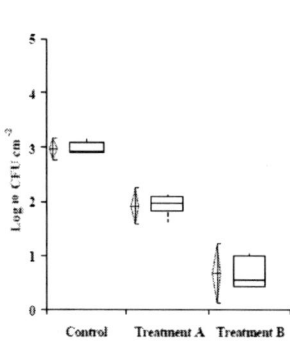

Conclusions

Despite advances in modern materials technology, many textiles are susceptible to spoilage by microbial growth when used in conditions that allow their growth. A wide range of biocidal treatments are available to prevent such spoilage and a wide range of tests exist that can be used to model such growth and determine the efficacy of preventative treatments.

In addition to protecting textiles in service, a number of biocidal treatments are now available that intend to introduce antimicrobial properties to textiles which bring hygienic or performance benefits. Although a number of standard methods exist, in general, these do not predict well the performance in service and fail to describe the benefit intended. It is however, possible to design study protocols that allow performance to be measured under realistic simulations, and support claims and illustrate benefits. With the current regulatory climate and with justified concerns over the exposure of humans to biocidal treatments (as well as the potential development of resistance by microorganisms to biocidal treatments and clinical antimicrobial agents), the need to understand the functionality of antimicrobial treatments and any benefits they may bring is becoming increasingly important.

References

1   Askew P D (2006), Biogenic Impact on Materials - Paper and Textiles - Springer Handbook of Materials Measurement Methods, Czichos H, Saito T, Smith L (Eds), Springer Velag, Germany.

2   Wypkema A W (2005), Microbicides for the protection of textiles, Directory of Microbicides for the Protection of Materials - A Handbook, Paulus W (Ed), Springer Verlag, Germany.

3   James AG, Hyliands D, Johnston H (2004), Generation of Volatile Fatty Acids by Axillary Bacteria, Int J Cosmetic Sci, 26, 3, 149 - 156

4   Rennie PR *et al* (1990), The Skin Microflora and the Formation of Human Axillary Odour, Int J Cosmetic Sci, 12, 5, 197 - 207

5   Marshall J,Holland KT, Gribbon EM (1988), A Comparative Study of the Cutaneous Microflora of Normal Feet with Low and High Levels of Odour, J Ap Microbiol, 65, 1, 61 - 68

6   Askew P D (2007), Exploring the Functionality of Antibacterial Plastics and Textiles, Antimicrobials in Plastic and Textiles Applications, PIRA Conference, Prague, Czech Republic, June 2007.

7   Anon (2008), Manual of Decisions for Implementation of Directive 98/8/EC Concerning the Placing on The Market of Biocidal Products.

ANTIMICROBIAL TEXTILE EFFECTS

Dr Heinz Katzenmeier
Sanitized AG
Lyssachstrasse 95, P.O. Box 1449, CH-3401 Burgdorf, Switzerland
Tel: +41 34 427 16 16   Fax: +41 34 427 16 19

BIOGRAPHICAL NOTE

Heinz Katzenmeier studied chemistry at the Technical University of Darmstadt where he finished his Ph.D. 1996 at the department of bioinorganic chemistry with his thesis "5,10,15,20-Tetrakis(4-phosphono-methylphenyl)-porphyrin, a new water-soluble porphyrin".

During a post doc stay at the Cilag AG Schaffhausen (CH), an affiliate of the Johnson & Johnson company he became familiar with the development of pharmaceutical active ingredients.

1998 he was moved to Zurich by Cilag AG to lead the Kilo-Lab of the Laboratory of Process Research at the University of Zurich, which is the competence-center of Cilag for the production of active pharmaceutical ingredients under GMP for first clinical trials.

In 2001 he moved to Clariant, Basel where he was responsible for projects with dispersed systems at pilot-plant scale level. Then in 2004 he built up the Clariant NanoTec Lab, an interdivisional competence centre for projects with small particles, dispersing and emulsification technologies.

Since February 2010 he became the head of Innovation and R&D at SANITIZED AG where he is responsible for the development of new biocidal products and evaluation of new concepts for antibacterial effects.

ABSTRACT

The antibacterial finishing of various textile fabrics is a still growing business. The fields of application range from apparel fashion for home-textiles to technical textiles.

Therefore biocides must be applied at a large number of different fabrics like cotton or synthetic materials by using different application-technologies like padding or exhaust methods.

The current major challenge is the application of biocides permanently on different fabrics to obtain a high wash fastness. This is relatively easy with fabrics which have reactive OH-groups, such as cotton. But with synthethic, mostly nonpolar materials which have no functional groups the fixation of a biocide on the surface of a fibre poses a significant problem.

Another challenge is the validation of the antimicrobial effect detected using standard microbiological tests and in particular the demonstration of the efficacy for the intended end uses.

Using as an example the product Sanitized® T 99-19 which is a biocide based on a quaternary ammonium compound, and having a functional trimethoxysilyl group, feasible solutions for these two problems are demonstrated.

This paper will show how products such as Sanitized® T 99-19 can be applied via both the exhaustion and padding processes which produce a wash fast antimicrobial finish. This wash fastness is subsequently tested by a preconditioning process with the antimicrobial efficacy evaluated using the JIS 1902-2002. Further to this a Phase 3 field trial using cotton towels is also described. Finally two odour inhibition studies are described, the first trial is the traditional olfactory panel evaluation, while the second employs the new electronic nose technology.

Antimicrobial Textile Effects

1. Introduction

Within Europe there is an annual demand for textiles in the range of > 200 Billion Euro. This demand is split into various parts like apparel-fashion, home-textiles or technical textiles.

Examples for applications in apparel fashion are sportswear, underwear, shirts, trousers und much more, very often finished with special properties like moisture-management, stain-release behaviour or hygienic finishes. Home-textile applications are the use for furniture, curtains or bedding. Especially for bedding hygienic properties like the protection against dust mites is required.

Examples for technical textiles are automotive applications, agriculture, architecture and much more. Also in the range of technical textiles special tailor-made properties are very important. This is especially important where textiles are used in an outdoor area. Here protection of the material against deterioration by microorganisms (bacteria, fungi) is absolutely indispensable.

Further to this today's consumers have high expectations where hygienic properties of textiles are concerned. In addition consumers are demanding textiles where the production of odour is minimised whilst the garments are being worn. As a result there is an increasing demand for textiles that are treated with a "hygienic finish". Therefore the market for a wide variety of fabrics that have been treated with an antibacterial finish continues to grow.

Another important trend is the ongoing development of household laundry being washed at low temperatures in order to save energy. Although this trend has positive benefits from an ecological viewpoint this is potentially likely to result in incomplete microbiological cleaning and therefore a poor hygienic outcome with clean clothes still being contaminated with microorganisms. Here the use of textiles treated with an antibacterial finish could make an important and sustainable contribution to the overall laundry hygiene process.

Due to the numerous and varied applications and uses, biocide treatments must be applied to a large number of different substrates such as natural materials such as cotton or synthetic materials such as polyester. There are also a variety of application-technologies for example: padding- or exhaust. The major requirement when applying biocides is to instil a degree of permanence onto the different fabrics therefore obtaining a high degree of wash fastness. This is relatively easy with fabrics that have reactive OH-groups such as cotton. However with synthethic materials it is a different story with most being non-polar and having no functional groups and therefore the fixation of a biocide to the surface of a fibre is frequently a problem.

A further challenge is the verification of the antimicrobial effect by means of standard microbiological tests and in particular the demonstration of the efficacy for the intended uses.

The antibacterial effect of a specific active substance or biocidal product can usually be verified by use of relatively simple standard microbiological test methods. However where odour production is concerned it is more complex. Evidence that an antibacterial finish which has shown inhibition or reduction of bacterial growth is frequently used to infer a reduction in odour. In the past odour changes were usually carried out by employing an experienced panel of people to make comments on the any odours detected. Now there is the possibility to use electronic nose technology which has the advantage of having a statistical validity.

Therefore it is essential to perform valid laboratory testing which simulates the end use conditions of an antimicrobial treated material allied to appropriate field testing.

This paper intends to use the product Sanitized® T 99-19 as a model of the processes described above. This biocidal product is based on a quaternary ammonium compound which has functional trimethoxysilyl groups.

2. Sanitized® Biocide T 99-19[1]

Sanitized® T 99-19 is a biocide, which can be used for the treatment of most commercially available fabrics. It is a quaternary ammonium-compound with trimethoxysilyl functionalities (figure 1).

Figure 1 – Active of Sanitized® T 99-19

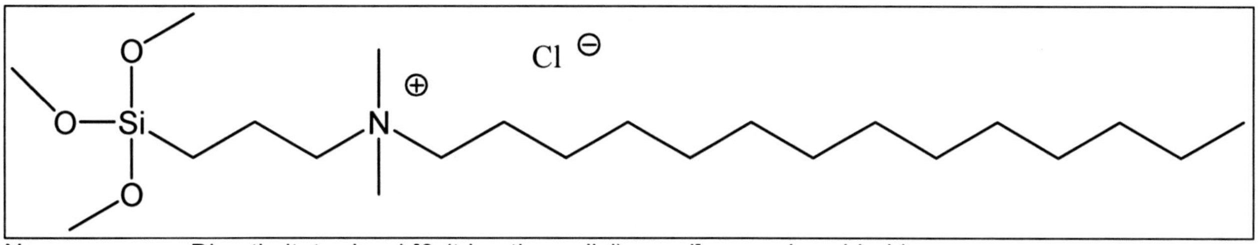

Name: Dimethyltetradecyl-[3-(trimethoxysilyl)propyl]ammoniumchloride
CAS: 41591-87-1
Formula: $C_{22}H_{50}ClNO_3Si$
Molecular mass: 440.18 g/mol

The concentrate version of Sanitized® T 99-19 contains 40 % of the active ingredient in triethylene-gycolmonomethylether (TGME).

In all following tests the diluted, 40% product is meant, if we talk about Sanitized® T 99-19.

3. Oligomerisation and binding to substrates of Sanitized® T 99-19

The trade form of the product Sanitized® T 99-19 contains 40 % of the active ingredient in triethylenegycolmonomethylether. In order to apply this product correctly it has to be diluted in water. It is known that in water a chemical oligomerisation reaction occurs, which leads to the formation of an oligomer with a chain length of approximately 6 monomeric units.

Due to the cationic nature of Sanitized® T 99-19 the molecule has a high affinity to anionic substrates with a high polarity such as cellulose.

Although the binding of Sanitized® T 99-19 to cellulosic substrates is relatively strong and consequently leads to finishes with a high wash fastness the affinity of the product to non-polar materials like polyester or polyamide is relatively weak and the preparation of antibacterial finishes with good wash fastness is a major challenge.

To meet this challenge SANITIZED developed an immobilisation and fixation technology for Sanitized® T 99-19 where it needed to be applied to non-polar textile substrates. This technology includes the use of a special crosslinker-system, Sanitized X-Linker PAD 26-19. This consists of a mixture of aluminium calcium and magnesium salts which are crosslink the Sanitized® T-99-19 molecule as shown in figure 2. This then enables a wash-permanent finish to be applied to even the most difficult of nonpolar substrates such as polyester or polyamide.

---

[1] The product T 99-19 is protected by a number of patents including: DE: 19928127, USA: 6376696, EP: 1194434, WO: 0078770, JP2003519090

Figure 2 – Crosslinking of Sanitized® T 99-19 by Sanitized X-Linker PAD 26-19

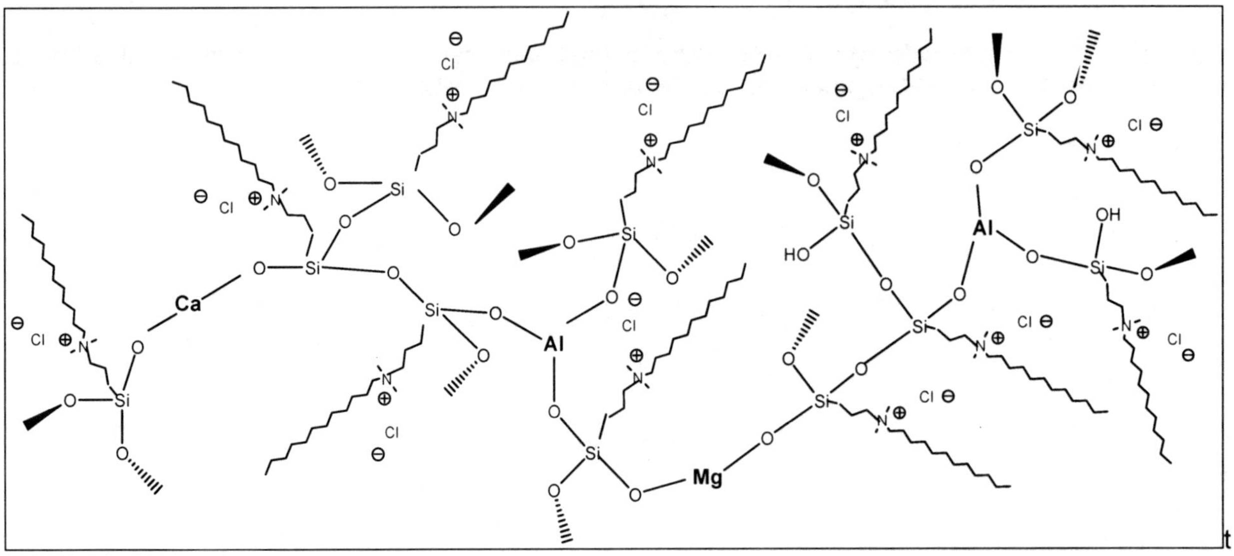

Note: In all the following studies, Sanitized® T 99-19 was applied to the polyester by means of Sanitized X-Linker PAD 26-19.

4. Antimicrobial efficacy (Minimal Inhibition Concentration)

The purpose of this work was to evaluate, and demonstrate the spectrum of antibacterial activity of Sanitized® T 99-19. The technique used was that of liquid cultures with Minimum Inhibitory Concentration (MIC)[2] of each test organism being determined visually at the end of the incubation period. Each MIC determination was repeated six times and the mean concentration of Sanitized® T 99-19 product calculated.

Table 1 – Minimal Inhibition Concentration (MIC) of Sanitized® T 99-19 against bacteria

Sanitized®T 99-19 Formulated Product (40%)	MIC
B. subtilis ATCC 6633:	26.0 ppm
E. coli ATCC 11229:	143.2 ppm
E. faecalis ATCC 51299:	71.6 ppm
K. pneumoniae ATCC 4352:	32.6 ppm
P. vulgaris ATCC 8427:	65.1 ppm
P. mirabilis ATCC 14153:	71.6 ppm
S. choleraesuis NCTC 6017:	247.4 ppm
S. aureus ATCC 6538:	32.6 ppm
S. aureus (MRSA) ATCC 33592:	17.9 ppm
S. epidermidis ATCC 12228:	17.9 ppm

The above MIC data indicates that Sanitized® T 99-19 has a wide spectrum of activity against bacteria with its intrinsic high efficacy especially against two major target species i.e. *S. aureus* and *K. pneumoniae*.

[2] MIC-method; SAN BIO – 21/95

Table 2 - Minimal Inhibition Concentration (MIC) of Sanitized® T 99-19 against fungi

Sanitized® T 99-19 Formulated Product (40%)	MIC
Candida albicans ATCC 10231:	313 ppm
Trichophyton mentagrophytes ATCC 9533:	52 ppm
Paecilomyces variotii ATCC 18502:	65 ppm
Penicillium funiculosum ATCC 11797:	521 ppm
Aspergillus repens DSM 62631:	469 ppm
Aspergillus fumigatus ATCC 9197:	156 ppm

MIC data against fungi indicates that Sanitized® T 99-19 has a broad spectrum of antifungal activity. Further to this the intrinsic efficacy of Sanitized® T 99-19 is especially high against *A. fumigatus* (the causative organism of farmer's lung), *P. varioti* which has known allergenic properties, and the dermatophyte species *T. mentagrophytes*.

Although these promising effects against fungi the remainder of this paper will be on the antibacterial effects. This is because where hygienic properties are concerned the major relevant target species are bacteria.

5. Microbiological Efficacy on Textiles

For the purpose of demonstration of the antimicrobial activity of the active substance Sanitized® T 99-19 against the bacteria *K. pneumoniae* (ATCC 4352), *S. aureus* (ATCC 33592 MRSA) and *S. aureus* (ATCC6538) when applied to the surface of the two major materials cotton and polyester an efficacy test was evaluated.

The technique used was that described in the Japanese standard JIS 1902-2002, but with a minor modification in the interpretation of the resultant data. In this the numbers of test bacteria were simply noted at the beginning and end of the incubation period and so a bactericidal or bacteriostatic effect was observed.

The results show that after 18 hours incubation, Sanitized® T 99-19 at concentrations of 0.4% and 1.2% on both cotton and polyester, there was a significant reduction in the numbers of viable bacteria.

So Sanitized® T 99-19 in a concentration of 0.4% and 1.2% has an excellent antibacterial efficacy against all test organisms (table 3). Therefore a concentration between 0.4 and 0.6% of Sanitized® T 99-19 is recommended for textile applications.

To confirm, that the recommended concentration range is correct and lower levels are not appropriate further testing was carried out at the lower concentration of 0.1 %. The results indicated that Sanitized® T 99-19 was not effective against *K. pneumoniae* (ATCC 4352) and this confirmed that the recommended concentration range was correct.

Further to this the untreated control test fabrics (No biocide) for both the 100% cotton and 100% polyester textiles showed significant growth of the test organisms.

Table 3 – Germ-reduction of T 99-19 at different textile substrates[3]

	Cotton, 100%			Polyester, 100%		
Concentration: Bacteria	0.1%	0.4%	1.2%	0.1%	0.4%	1.2%
K. pneumoniae (ATCC 4352),	no reduction	Good log 4.7	Good log 4.7	no reduction	Good log 4.7	Good log 4.7
S. aureus (ATCC 33592 MRSA)	Good log 3.2	Good log 3.2	Good log 3.2	Good log 3.2	Good log 3.2	Good log 3.2
S. aureus (ATCC6538)	Good log 3.5	Good log 3.5	Good log 3.5	Good log 2.5	Good log 2.5	Good log 3.5

A log reduction of >1 log-unit indicates good performance (unwashed textile)

6. Determination of the wash-permanency of T 99-19

The purpose of this "in use" study was to evaluate, and demonstrate the antimicrobial activity of the formulated product Sanitized® T 99-19 against the different bacteria when applied to the surface of two major materials i.e. cotton and polyester after washing.

The technique used was that described in the Japanese standard JIS 1902-2002, but with a minor modification in the interpretation of the resultant data. Further to this there was a preconditioning step of washing twenty times according to EN ISO 6330.

The numbers of test bacteria were simply noted at the beginning and end of the incubation period to enable a sustained bactericidal or bacteriostatic effect to be observed.

A log reduction of > 0.6 log-units indicates a good performance (washed textile)

The test showed, that after twenty washes at 40°C Sanitized® T 99-19 at a concentration of 1.2%, following a preconditioning treatment of twenty wash cycles at 40°C, showed excellent antibacterial efficacy against all test organisms at all substrates (table 4).

Furthermore the test showed that after twenty washes at 40°C at a concentration of 0.4% the test fabrics treated with Sanitized® T 99-19 following an 18 hour incubation period lower numbers of viable bacteria were detected.

---

[3] method JIS L 1902 – 2002

Table 4 – Germ-reduction of T 99-19 at different textile substrates after 20 washes[3]

	Cotton, 100%			Polyester, 100%		
Concentration: Bacteria	0.1%	0.4%	1.2%	0.1%	0.4%	1.2%
K. pneumoniae (ATCC 4352),	no reduction	Good log 0.75	Good log 1.6	no reduction	low reduction log 0.34	Good log 1.4
S. aureus (ATCC 33592 MRSA)	Good log 2.0	Good log 2.3	Good log 3.2	no reduction	Good log 2.5	Good log 3.2
S. aureus (ATCC6538)	Good log 1.6	Good log 3.5	Good log 3.5	no reduction	Good log 1.1	Good log 3.5

7. Field study: Antibacterial Effect "in-use"

The purpose of this work was to evaluate, and demonstrate the antibacterial efficacy of Sanitized® T 99-19 under relevant use conditions.

The study employed cotton towels cut lengthwise down the middle with one half treated with Sanitized® T 99-19 and other half untreated. Both the treated and untreated half towels were washed separately on twenty occasions, after which the treated and untreated halves were sewn back together. Following this the towels were placed at suitable representative locations, such as the cafeteria and restrooms, in SANITIZED AG's company headquarters building and used over a one week period.

Immediately after use the towels were taken to the microbiology laboratory and paired samples cut from the same position on the treated and untreated side of each of the towels. Bacteria were removed from the surface of the towel samples by shaking in a liquid bacterial culture medium. The numbers of bacteria extracted were then determined by a standard plate count technique. Data obtained from the paired samples obtained for each towel was used to calculate the efficacy of the antibacterial treatment

The test was designed to be a realistic evaluation of typical towel usage. The 20x preconditioning washing process would reflect the efficacy of older towels. Also it is reasonably typical for towels to be changed on a weekly basis, and hence the one week test period.

Given the preconditioning washes and the week long test it was considered that the test conditions would reflect typical usage meaning that the towels were in contact with for example: dirt, skin particles, fat, remaining soap and bacteria. The isolation of bacteria took place immediately after the collection of the towels from their locations. There were no additional controls, or other factors such as antibacterial soaps which could influence the results.

The test was blind in that only the two people within the SANITIZED AG headquarters building knew the identification key. Further to this neither of these two members of staff were part of the microbiology group who analysed the samples for the presence of bacteria.

The test showed, that there was an average reduction of the number of colony forming units on the towels after 20x pre-washing was a factor 2.0 (figure 3). This is a significant reduction.

Figure 3 - Log.-reduction of bacterial growth at the "in use"field-study

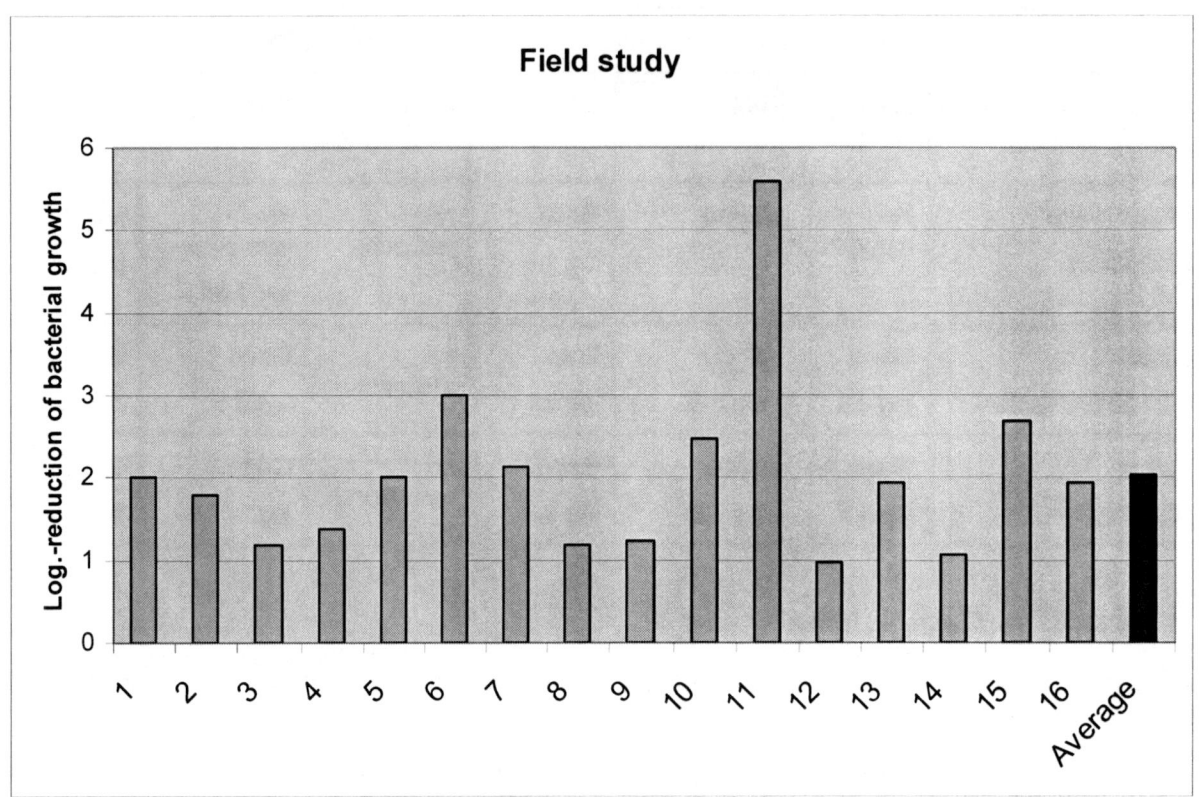

8. Field study odour-reduction

Purpose of the study was to find out, if the inhibition of the bacterial growth at textiles, which are finished with Sanitized® T 99-19 correlates with a consequent reduction of odour production. To measure the odour production under real wear conditions a field study was instigated through the Hohenstein Institute[4].

The aims of the test were as follows:

I	**Determination of the total antibacterial activity of the textile material**
II	**Investigations in real wear situation (athletic activity)**
IIa	**Evaluation of sweat odour reducing ability in vivo of textile material**
IIb	**Determination of the total germ count on textile material after athletic activity**

The test-material was a standard polyester (100% PES), finished with 1.0% of Sanitized® T 99-19.

Part I "Determination of the total antibacterial activity of the textile material

The test was carried out according to the requirements of DIN EN ISO 20743A: 10-2007 „textiles – determination of the antibacterial activity of antibacterial finished products".

The evaluation of the total antibacterial activity of the textile material indicated, that there was significant antibacterial activity against *Staphylococcus. aureus* (ATCC6538) with a log reduction of 2.92 being recorded. There was even greater antibacterial activity against the second test organism *Klebsiella pneumoniae* (ATCC 4352) with a log reduction of 4.47 being recorded.

---

[4] The Hohenstein Institute, Institute for Hygiene & Biotechnology, Schloss Hohenstein, 74357 Bönnigheim, Germany

Part II "Investigations in real wear situation (athletic activity)"

Swatches of the samples were attached to the axilla of the test persons in a randomised manner on each side of the body. Each test person then carried out intensive athletic activity for 30 minutes. All test persons reached their aerobic threshold moving into the anaerobic zone with 75-85% of their maximum heart rate. At this heart rate, carbohydrates and fats are burned for energy production in the muscle cells.

After this phase of the test and the textile swatches were soaked with sweat, they were processed according to VDA 270: "Determination of the odour characteristics of trim materials in motor vehicles". Modifications were made regarding temperature, sample size, time of storage and storage container size.

Part IIa "Evaluation of sweat odour reducing ability in vivo of textile material"

This method was developed by the Hohenstein Institute and is used to assess the odour concentration above odour thresholds typical in humans, post exercise.

Odour intensity was judged on the basis of VDI 3882 Part 1: "Olfactometry – Determination of Odour Intensity". The number of judging panellists was four. Panellists have been trained according to DIN EN 13725: 2003 "Air quality – determination of odour concentration by dynamic olfactometry".

The value of odour intensity was assessed after one hour by comparing the untreated material with the treated material after following category scaling (table 5).

Table 5 – Odour Intensity Scale

Odour	Intensity level
Extremely strong	6
Very strong	5
Strong	4
Distinct	3
Weak	2
Very weak	1
Not perceptible	0

The values of the pair wise odour intensity estimations were summated for an overall statement. In addition odour estimations of male and female test persons are displayed separately. Odour estimation of the untreated material was set to 100%.

The results of the study are summarised in the following table:

Table 6 - Odour Reduction Differences: Males and Females

	Untreated material	Treated material	Difference*
Male	52%	41%	11%
Female	48%	41%	7%
Total	100%	82%	18%

*Difference between treated and untreated material

Results of the odour analysis by the panellists showed a basic odour burden of the treated and untreated material. With a reduction of 18% the overall odour impression was significantly lower for the treated material compared with that noted for the untreated material (figure 4).

Figure 4 - Odour-reduction (all test persons)

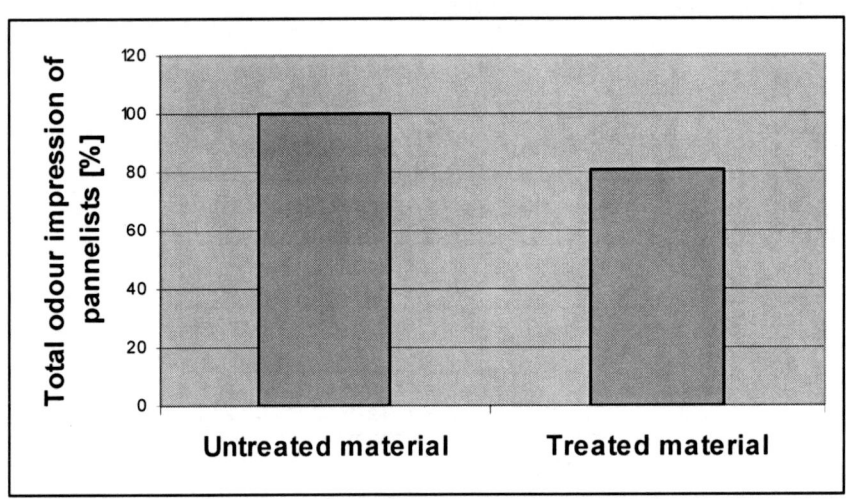

Further to this odour from the males (figure 5) was considered by panellists to be more intense when compared to that of females (figure 6). The difference in odour intensity between treated and untreated material was more distinct in males.

Figure 5 – Chart odour reduction, male test persons

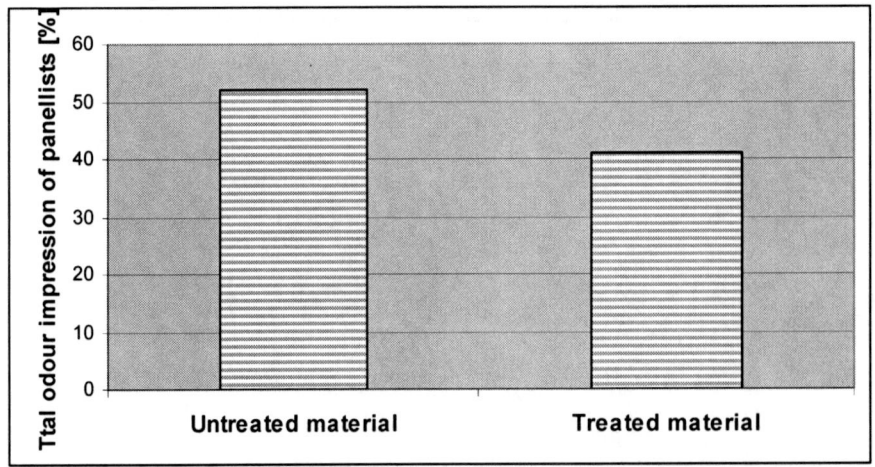

Figure 6 – Chart odour reduction, female test persons

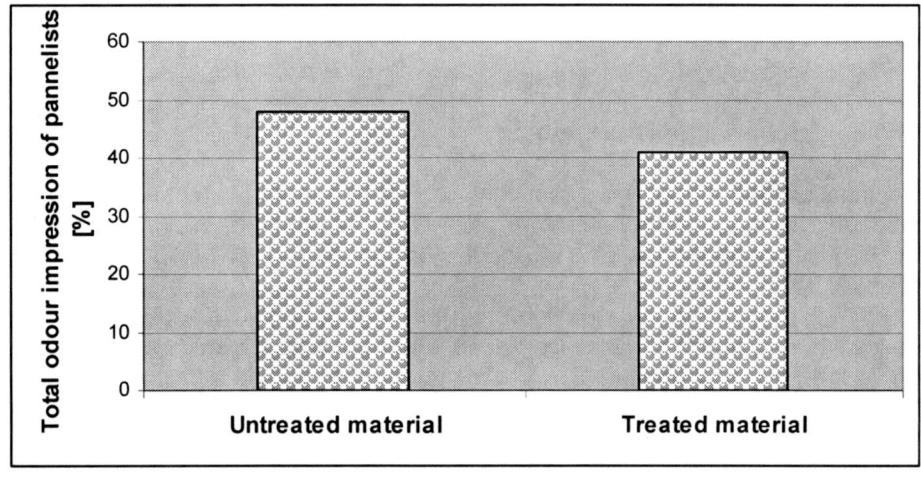

In this evaluation of odour reducing ability of Sanitized® T 99-19 on treated textile swatches the sweat and bacteria are transferred onto the textile swatches from the skin. Thus an effective antibacterial finished textile material can have an odour reducing effect by reducing the bioburden on the textile. This is especially important where damp clothes are stored for some time prior to washing.

Odour production that takes place directly on the skin is not likely to be affected.

***Part IIb* "Determination of the total bacterial count on textile material after athletic activity"**

Determination of total bacterial count on textiles following DIN EN ISO 11737-1: 2006: "Sterilisation of medical devices – microbiological methods – Part1: determination of a population of microorganisms on products".

After sweat inoculation (due to sport activity) there was a reduced bacterial count on the treated material with 17 of the 20 test persons tested. The average reduction in bacterial count on the treated material after the real wear sweat test was 80% (figure 7).

Figure 7 – Chart: Reduction of germ count of sweat inoculated textile material

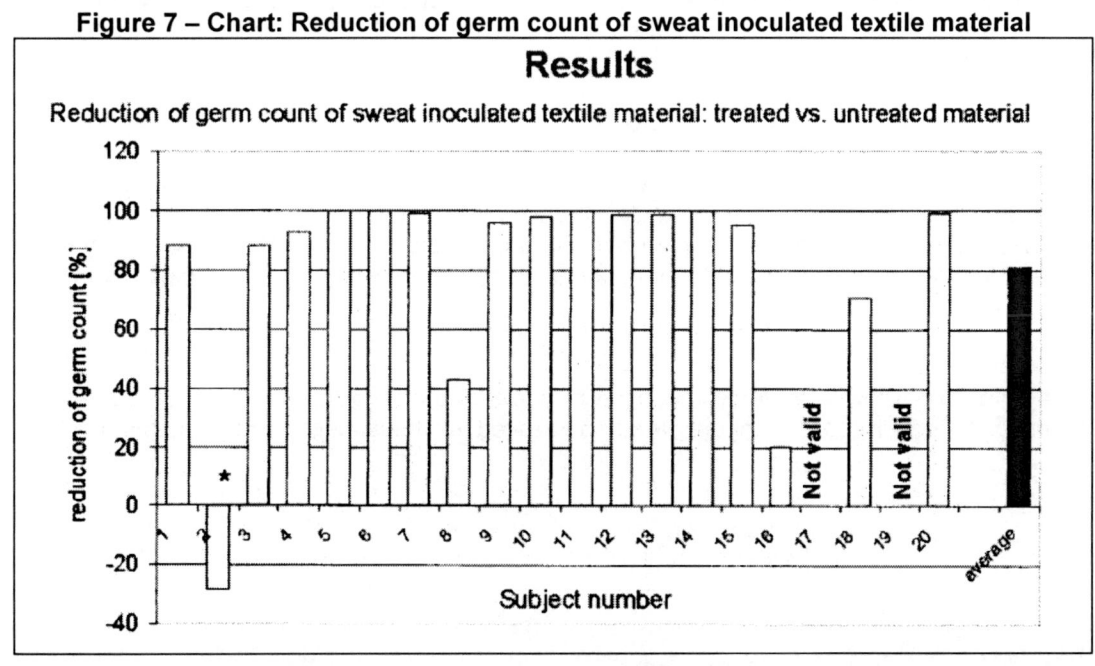

As anticipated the total bacterial count found on the textile varied between the test personnel which was due to each individuals bioload. Thus the effectiveness of the treated textile varied between each of the people taking part. These variations were expected and were within anticipated limits.

Finally odour production of individuals can vary widely and it is subjective so it was decided to investigate another methodology to try to overcome these deficiencies. Therefore a second method based use of an electronic sensor was investigated.

9. Odour inhibition – Electronic Nose Technology

The aim of the evaluation was to perform odour tests on anti-microbial fabrics with T 99-19 treated cotton and polyester materials using a highly quantitative technology, which is independent from human variations and not based on olfactory detection of the odour. Further to this given its numerous readings it is considered statistically valid.

Therefore the following test using a new high quantitative electronic nose was performed by Scensive Technologies Ltd[5].

Two sets of samples were tested:

1)   Cotton Twill swatches ($200g/m^2$), approximately A4 size:
a)   Unwashed samples – Untreated and Treated with 0.1%, 0.4% and 1.2% of T 99-19 biocide (3off A4 swatches each)

2)   Polyester (Trevira, $220g/m^2$) swatches approximately A4 size:
a)   Unwashed samples – Untreated and Treated with 0.1%, 0.4%. 0,8% and 1.2% of T 99-19 biocide (3off A4 swatches each)

Test Details

Anti-microbial testing was undertaken using the Standard Method AM802, designed to simultaneously measure odour reduction capabilities of fabrics and relate these to the anti-microbial effect of a biocide. Each set of tests included 'blank' or zero controls as well as positive controls.

The micro-organism used was *Staphylococcus aureus* NCTC 7447 (ATCC 6538) which is the micro-organism used in the standard AATCC 100 test for anti-microbial activity.

For the test to the sterilized textile samples a certain amount of a bacterial culture is added and then the samples are sealed in PVC-bags. The sealed samples are incubated at 37°C in the PVC sachets overnight for a minimum of 16 hours.

Then the odour inside the PVC-bags is measured by the Bloodhound® Odour Detector.

Plate counting was performed to check on bacterial kill or viable load, using the methodology given in the AATCC 100 test protocol, where 20ml of sterile 3%v/v Tween detergent (Tween 80) is introduced directly into the PVC sachets to stop bacterial growth and extract the bacteria from the fabric. Serial dilutions into sterile Tween 80 or LB broth may be made for actual plate counting, using 1ml aliquots per plate.

Results

The results for **cotton** showed a strong reduction of odour in the range of about 90%. The following chart (figure 8) shows the odour-exposure of untreated and treated textile-samples in case of cotton.

---

[5] Scensive Technologies Limited, Metic House, Ripley Drive, Normanton, West Yorkshire, WF6 1QT, United Kingdom

Figure 8 – Odour-exposure at cotton-substrate

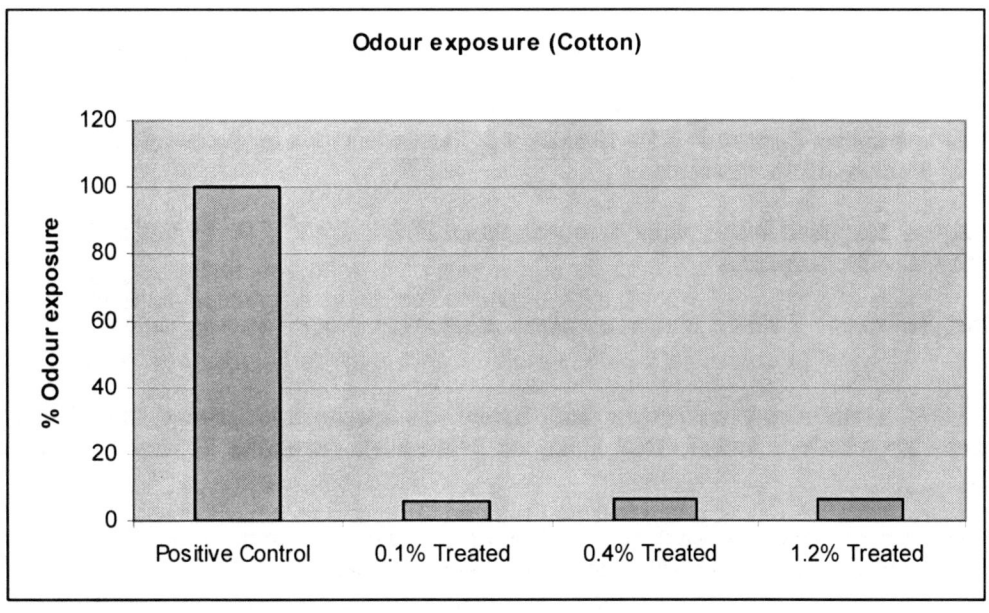

Odour-exposure was calculated by subtracting odour-reduction from 100% (no odour-reduction).

The results for **polyester** also showed a strong reduction of odour in the range of about 85%. The following chart (figure 9) shows the odour-exposure of untreated and treated textile-samples in case of polyester.

Figure 9 – Odour-exposure at polyester-substrate

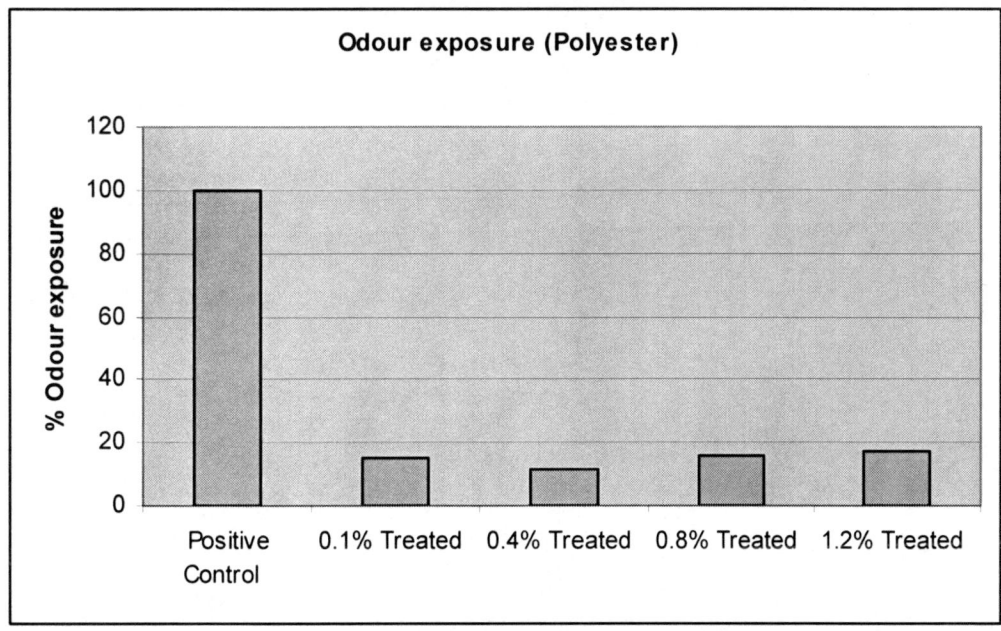

Odour-exposure was calculated by subtracting odour-reduction from 100% (no odour-reduction).

From the above results it is concluded that the antimicrobial finish provided by Sanitized® T 99-19 was effective in reducing odour under the conditions of the test at cotton as well as at polyester fabrics. This was confirmed by the bacterial counts following culturing onto culture media and overnight incubation.

9. Summary

This paper through the use of Sanitized's biocidal product T 99-19 demonstrates the value of the use of antimicrobial finishes applied to textiles.

The data presented shows Sanitized® T 99-19 to be a product which has an excellent biocidal activity against a wide range of problem microorganisms.

The testing regime described in this paper demonstrates that Sanitized® T 99-19 has a high level of efficacy when applied as a finish to textiles.

Further to this Sanitized® T 99-19 shows excellent application properties with natural materials such as cotton.

Sanitized® T 99-19 when used in conjunction with Sanitized's special crosslinker-system, X-Linker PAD 26-19, can provide an excellent antimicrobial finish on problematic non-polar textiles such as polyester or polyamide.

Finally when Sanitized® T 99-19 is applied to textiles as a hygienic finish the end result is proven efficacy, easy applicability which is associated with a high wash-permanency.

For further information please feel free to contact SANITIZED AG[6]

---

[6] **SANITIZED AG,** Lyssachstrasse 95, P.O. Box 1449, CH-3401 Burgdorf
T: +41 34 427 16 16 | F: +41 34 427 16 19

A DURABLE, LIGHT AND HEAT STABLE AND EASY-TO-USE ANTIMICROBIAL PRODUCT FOR TEXTILES AND NON-WOVENS

Tirthankar Ghosh, Ph.D. Dow Microbial Control
The Dow Chemical Company
727 Norristown Road, Spring House, PA 19477, USA
Tel: 001 215-619-5410 Fax: 001 215-619-1654 email: tghosh@dow.com

BIOGRAPHICAL NOTE

Tirthankar (Tutul) Ghosh received his Ph.D. in Organic Chemistry from Texas Christian University, Fort Worth, Texas, following his M.S. in Chemistry from Jadavpur University, Kolkata, India. Prior to joining the Dow Chemical Company, he carried post-doctoral research at Michigan State University and Princeton University. Tutul has published over 15 articles and holds 15 granted US patents. Currently Tutul is a Distinguished Scientist in the Dow Microbial Control division. He is currently involved in the development and commercialization of a novel silver antibacterial technology.   He is also leading the effort to look beyond the current product portfolio in the hygiene market segment, to identify and lead the exploration for the development of the next generation of products.

ABSTRACT

Recently Dow Chemical Company has introduced SILVADUR™ ET antimicrobial, a silver based technology that is able to provide a durable, light and heat stable antimicrobial performance to a wide range of textiles and nonwoven materials. Unlike other products in the market today, SilvaDur™ ET is a liquid delivery system for silver and is completely particle free. It utilizes a proprietary polymer technology to provide a durable antimicrobial finish with minimal effect on the feel of the fabric. The liquid product is readily water dilutable, allowing formation of phase stable and solids free finishing bath solutions for easy and rapid processing of fabric. Extensive evaluation has shown that the product is compatible with a wide range of natural and synthetic fabrics and provides durable (>50 launderings) antimicrobial performance.

Introduction

The battle to control and eliminate harmful organisms, before they can contaminate our food, render our water undrinkable and enter our bodies causing a wide variety of diseases, has been a long fought one and the fight continues. Over the years antimicrobial materials have provided the first line of defense. These materials range from organic compounds (e.g isothiazolones, 3-iodopropynylbutylcarbamate, 2,4,4'-trichloro-2'-hydroxydiphenyl ether, quaternary ammonium compounds) to metals and their salts (e.g. arsenic, tin, copper, zinc, and silver). Some organic biocides have been deemed unsafe for use in medical and food-contact applications due to their inherent toxicity and the increased risk of exposure when the biocide leaches from the treated article.   Organic biocides are also unstable at high temperatures, with most degrading at temperatures exceeding 150 degrees Celsius. Of the metals, arsenic has been used widely to treat wood but recently there has been a voluntary withdrawal of CCA (Chrome-copper-arsenate) treated wood because of mammalian toxicity. Tin is another biocide that has been used for the past few decades as a very effective marine antifoulant. However, recently the International Maritime organization (IMO) has passed legislation to ban the use of tin based marine antifoulant paints because its toxicity towards marine organisms.[i] In this regard silver is a very effective antimicrobial agent with low toxicity.

Silver has been used as an antibacterial agent for many centuries (Fig 1). Ancient Greeks and Romans used silver coins to preserve water and other liquids. Even today NASA uses metallic silver as a source of Ag^+ ions to maintain water purity in the space shuttle. Silver also is commonly used in medicinal uses in eye drops, nasal spray and burn ointments, and is beginning to replace chlorine, now suspected to have long-term toxic effects, in water filtration systems for hospitals, apartments, pools, schools and municipalities.

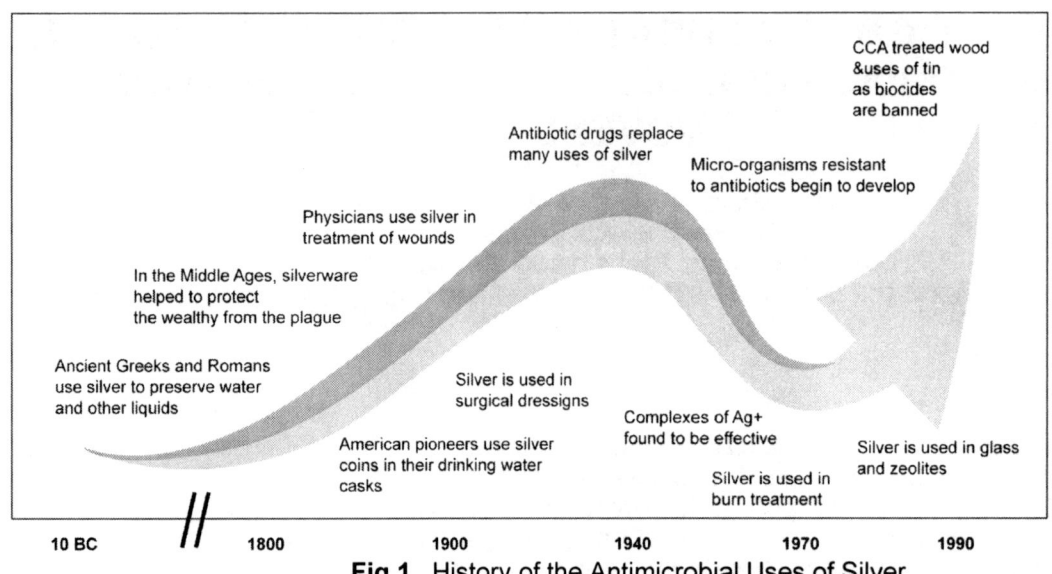

Fig 1. History of the Antimicrobial Uses of Silver

Silver exerts its antimicrobial effect at very low doses and is referred to as "oligodynamic" which means "active with few". This terminology was coined by von Naegelli in 1893.[ii] It is well established that free silver ion (Ag^+) is the active antimicrobial agent. Ag^+ is adsorbed upon contact with negatively charged bacterial cell surfaces. The inhibitory action of Ag^+ is due to strong interaction with thiol groups located in cell respiratory enzymes in the bacterial cell.[iii] In addition silver ions have been shown to interact with structural proteins and bind preferentially with DNA bases to inhibit replication. It has also been shown that silver ions have no effect on human cells in vivo. In order to achieve a bactericidal effect, Ag^+ must be in solution at the bacterial cell surface.

Based on this understanding of mode-of-action, a soluble source of Ag^+ (e.g. silver nitrate) should be the ideal bactericidal agent. Unfortunately solutions of Ag^+ are very photosensitive and are readily reduced to metallic silver when exposed to light, causing reduced activity and staining. In addition, soluble Ag^+ salts are easily removed by washing and cannot therefore provide a long term antibacterial effect.

Background

In modern times, the use of silver as an antimicrobial agent was first adopted in the plastics industry. About 20 years ago the industry became very interested in providing a variety of antimicrobial surfaces, e.g. kitchen counters, toys, etc to prevent the spread of germs. In order to incorporate Ag^+ in plastics the industry has provided silver delivery systems that are stable to heat and can withstand the extrusion process.

These systems to date have focused on three approaches. The first approach utilizes encapsulating the Ag^+ in an inert matrix, such as soluble glass, with complete dependence on leaching of Ag^+ to provide antibacterial activity.[iv] The release of silver from the glass matrix is diffusion controlled; over time there will be a build up of a high level of silver ions in the matrix. This ultimately will result in discoloration. A second approach is based on incorporating the Ag^+ into a zeolite matrix by ion exchange.[v] In this case, the release of Ag^+ is mediated by ion exchange with sodium or potassium ions present in the environment. In aqueous systems, the release of Ag^+ is controlled by the concentration of ions (e.g. Na^+) and humidity level in the surrounding environment which does not necessarily correlate to the level of bacterial contamination. The third approach has been to use insoluble silver salts (e.g. silver chloride, silver sulfate, silver zirconium phosphate,[vi] etc.) and then rely on the solubility constant (K_{sp}) to release Ag^+. All the products in this category are solid materials, which limits their applicability due to formulation difficulty.

In addition, an overarching factor in the solid delivery systems is that of particle size. At smaller silver carrier particle size, higher levels of silver ions are available on the surface which results in higher antibacterial activity. These high release rates of silver ions are not beneficial in this type of diffusion controlled system, as high levels of free silver ions will result in discoloration and reduced activity over time.

The last few years has seen a tremendous growth of silver in textiles and nonwovens to provide for odor control and antibacterial activity. There are two ways to introduce silver. It can be incorporated in the fiber (expensive) or applied as a coating later during the finishing process. All the solid materials discussed above are not suitable to be applied as a coating via an aqueous treatment bath. We realized that a viable delivery system of silver ions has to overcome the following performance deficiencies:

- Non darkening on exposure to light and heat
- Release of silver ion is not controlled by the size of the delivery system
- Liquid, preferentially water based that can be easily diluted in anionic, cationic or nonionic matrices
- Controlled release of Ag^+ based on demand

This paper will describe the development and performance of a durable, light and heat stable and easy-to-use antimicrobial finish for textiles and nonwovens

Our Approach

We believe that the most effective pathway to overcome the first two technical barriers listed above is by using a different mechanism to stabilize and release Ag^+. Instead of using diffusion or ion exchange we will use metal-ligand complexation chemistry to stabilize Ag^+ (Eq 1).

$$Ag^+ X^- + nL \rightleftharpoons \left[AgL_n \right]^+ X^- \quad (Eq\ 1)$$

The reverse reaction in Eq. 1 can be exploited to release Ag+ from the complex. Ligand choice is important to enable tuning of this silver release equlibrium. (Fig 2). The ligand is incorporated into a polymer with extensive architecture. The role of the polymer will be to disperse the complex in water and help in compatibility and durability of the complex with the article of use.

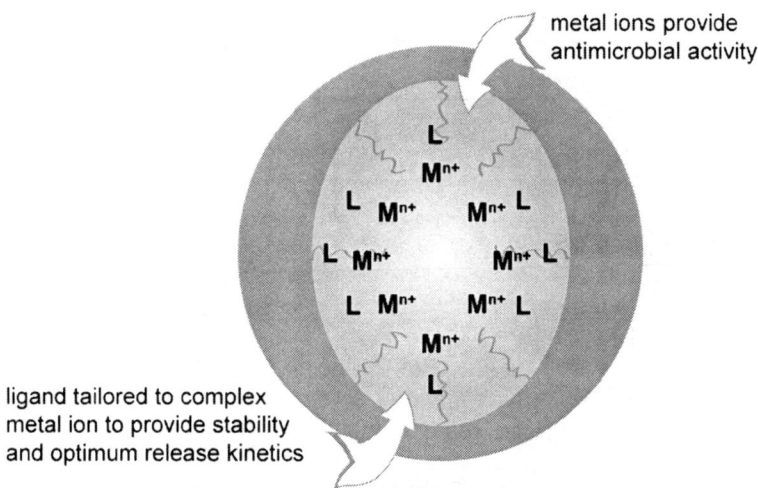

Fig 2. Delivery system for Ag^+ using polymeric ligand architecture

Ligands for binding Ag(I)

Nomiya and co-workers have shown that a wide variety of ligands form complexes with Ag^+.[vii] These complexes have been shown to have good antibacterial activity and do not discolor upon exposure to light. It is also encouraging to see that some of the complexes have increased activity against bacteria and fungi compared to $AgNO_3$ (Table 1).

Table 1. Minimum inhibitory concentration (MIC) for Ag^+ complexes

Complex	S. aureus[1] (ppm)	P. aeruginosa[2] (ppm)	A. niger[3] (ppm)
AgNO3	>1600	6	>1600
[Ag(imidazole)2]NO3	15.7	7.9	15.7
[Ag(1,2,4-triazole]	125	7.9	1000
Ag(RS-Hpyrrld)	31.3	15.7	15.7

[1]*Staphylococcus aureus* is a gram positive bacteria
[2]*Pseudomonas aureginosa* is a gram negative bacteria
[3]*Aspergillus niger* is a fungus

Imidazole   1,2,4-Triazole   2-Pyrrilidone-5-carboxylic
                                       acid (Hpyrrld)

Although the authors have shown that Ag$^+$ complexes with heterocyclic moieties as ligands have intrinsic antibacterial and antifungal activity, there is no data in the literature indicating that this activity can be realized when the ligand is incorporated in a polymer matrix.   We believe that by attaching these Ag$^+$ complexes to the right polymer architecture it should be possible to design an innovative approach to solve the key technical challenges outlined earlier.

SILVADUR™ ET

The Dow Chemical Company's silver based antimicrobial product (SILVADUR™ ET) is a water dilutable liquid formulation that addresses all the shortcomings of current silver antimicrobial products targeted for the textile and nonwoven industries.   SILVADUR™ ET contains a proprietary polymer and silver in a water/ethanol solution.   It is completely soluble in textile treatments baths, but forms an an insoluble interpenetrating polymer network with Ag$^+$ upon application to a fabric and drying.   The formation of this insoluble network results in a very durable antimicrobial finish.   This process is represented graphically in Fig 3.

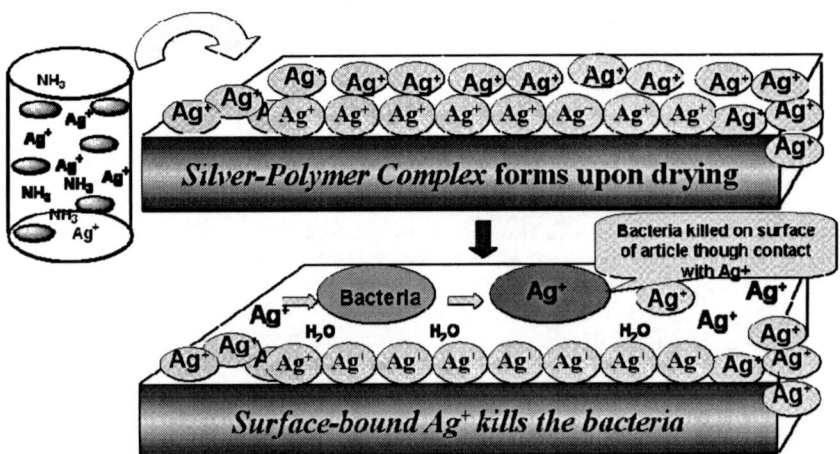

Fig 3. Formation of a Ag$^+$-polymer complex

When organisms land on the surface of the treated fabric (Fig 3) the free Ag$^+$ interacts with the organism resulting in cell death.   As the initial available Ag$^+$ is used up by interaction with organisms, more Ag$^+$ is released from the complex (K_{eq}) and the process continues.

The excellent antimicrobial efficacy of SILVADUR™ ET can be demonstrated by the determination of the Minimum Inhibitory Concentration (MIC) versus a number of test organisms including gram positive gram negative bacteria and fungi.   The results are shown in Table 2.

Table 2. MIC for SILVADUR™ ET (ppm of silver)

S. aureus	E. coli	K. pneumoniae	P. aeruginosa	C. ammoniagens	A. niger	C. albicans
0.43	0.56	0.13	0.11	4.25	5.0	5.25

Application and Performance

Pad application

SILVADUR™ ET readily dissolves in water to provide a light stable homogenous treatment solution which contains no particulate matter. Using the wet pickup (WPU) of the fabric the concentration of SILVADUR™ ET in the bath can be adjusted to attain the Ag^+ level required (note: SILVADUR™ ET contains 3% silver). One is easily able to obtain a homogenous antibacterial finish with a silver concentration very close to that targeted because of the phase stability and complete solubility of the active material in the treatment bath. Other additives (e.g. oil repellents, softeners) can also be added to the SILVADUR™ ET containing treatment bath, the only caveat being that the additives should be stable at pH 9.

Nonwovens

To determine the effective dose of silver needed to protect a nonwoven article, SILVADUR™ ET was applied topically as both a stand-alone treatment and in conjunction with a binder as might be used in resin-bonded nonwoven production. Two different webstocks (15 gsm 100% polypropylene spunbond and 35 gsm 100% polyester pointbond) were selected to represent systems employed in both consumer and industrial applications.

As a stand-alone treatment, the SILVADUR™ ET was applied to the web using a pad-saturation process. The nonwoven web was passed through a bath of standard tap water and weighed to measure the wet pickup (WPU) specific to the web. Based on this WPU, the SILVADUR™ ET was added to tap water adjusted at a pH of 9.5 to obtain the desired bath concentration. For example, a WPU rate of 300% and a target of 100 ppm silver on the dried web would require 33 ppm of silver in the bath solution. The webstock was passed through the bath and then dried 2 minutes at 149°C. The silver level was determined by inductively coupled plasma atomic emission spectroscopy (ICP-AES) and antibacterial activity was evaluated using the AATCC Method 100 test as shown in Table 3.

Table 3. Efficacy of SILVADUR™ ET on spunbonded polypropylene

Ag (ppm)	AATCC Method 100 Log Reduction/24 h contact time	
	S. aureus	*K. pneumoniae*
0	0	0
85	2.4	3.0
100	2.4	3.0
125	4.7	3.8
200	>5.4	3.7

The treated fabric did not show any discoloration even at the highest silver treatment. As the data is Table 3 suggests, even at the lowest tested loading level of 85 ppm the fabric demonstrated excellent antibacterial activity.

To measure the effect of a latex binder on the antimicrobial and fungal performance, a web of 35 gsm 100% PET was treated (as described above) with a solution of SILVADUR™ ET and a semi-durable binder. Again the silver level was determined by ICP-AES and antibacterial activity was evaluated using the AATCC Method 100 test. Antifungal performance was evaluated by using the AATCC Method 30. Results are shown in Table 4.

Table 4. Efficacy of SILVADUR™ ET in Combination with a semi-durable acrylic binder

Ag (ppm)	AATCC Method 100 Log Reduction/24 h contact		AATCC Method 30
	S. aureus	*K. pneumoniae*	*A.niger*
0	0	0	Low Growth
90	3	3	No Growth
180	3	3	No Growth
400	4	4	No Growth

The data in Table 4 clearly indicate that the presence of an acrylic binder does not alter the performance of SILVADUR™ ET. Excellent antibacterial activity is observed at even at the lowest testing loading level of 90 ppm. Antifungal activity of silver is not typical, particularly for silver nitrate.[viii] The data in Table 2 suggests that SILVADUR™ ET provides good antifungal activity.

To demonstrate that SILVADUR™ ET treatment does not have any detrimental effect on the physical properties of the nonwoven fabric, the samples were subjected to a battery of testing to measure tensile strength, color and stiffness (or hand).

Tensile strength was measured per ASTM D882-02 using a 2-inch gap setting and 12-inch per minute crosshead speed.[ix] Due to the anisotropic nature of the web, a minimum of eight specimens in each direction was tested dry and wet. For the wet systems, the specimens were immersed in the noted solvent for 30 minutes, removed, blotted, and tested immediately. Results are shown in Table 5.

Table 5. Tensile Strength of 100% PET treated with SILVADUR™ ET

| | Tensile Strength (g) | | | |
	Ag (0 ppm)	Ag (90 pm)	Ag (180 ppm)	Ag (400 ppm)
MD –Dry	4041	4724	4767	4599
Wet-H_2O	1317	1317	1317	1226
Wet-IPA	363	454	454	595
CD –Dry	817	863	999	849
Wet-H_2O	318	318	363	241
Wet-IPA	91	91	91	109

Outside of typical test variation, the SILVADUR™ ET treatment was found to have no impact on the strength of the acrylic binder. The machine direction (MD) tensile shows a slight increase in dry tensile, but no appreciable increase in wet tensile with either water or isopropyl alcohol as solvent. The cross directional (CD) tensile are virtually unchanged. Overall SILVADUR™ ET treatment does not alter the beneficial tensile properties of a semi-durable acrylic binder.

Color was measured using a Hunter L*a*b spectrophotometer on four plies positioned against a standardized black backing. The color is read on a minimum of three separate areas across both the web face and back. Stiffness was measured using a Thwing-Albert Handle-Ometer. Sample size is 4in² with a 10mm gap setting. These results are presented in Table 6 below.

Table 6. Color and Stiffness of 100% PET treated with SILVADUR™ ET

| Sample | Color | | | Stiffness |
	L	a	b	(g)
Control (100% PET)	86.13	-3.88	+1.80	14.0
100% PET + binder	87.71	-4.01	+2.02	16.9
100% PET + binder + 90 ppm silver	86.55	-3.92	+1.87	17.9
100% PET + binder + 180 ppm silver	86.58	-3.91	+1.80	19.0
100% PET + binder + 400 ppm silver	86.41	-4.15	+2.87	20.0

A slight development of yellow color was observed for the fabric containing 400 ppm of silver. At the effective dose level of 90 ppm and 180 ppm, there was no discoloration compared to the untreated control.

The quantitative measurement of stiffness shows no additional stiffness or softening of the treated article above that contributed by the semi-durable binder. This data shows that SILVADUR™ ET can be blended with a semi-durable acrylic binder, applied by conventional means to any substrate. The treatment does not affect tensile, color, or stiffness.

Textiles

To demonstrate the durability of SILVADUR™ ET, 100% cotton twill was treated with SILVADUR™ ET using the standard padding and drying operation. The samples were then subjected to a number of home laundering cycles using a Launder-Ometer® and the AATCC method 61 type No. 2A. After the washing cycles the samples were analyzed for silver content and antibacterial efficacy measured by AATCC Method 100. The results are tabulated in Table 7.

Table 7. Durability of SILVADUR™ ET on 100% cotton twill

Treatment	Wash Cycles	Silver (ppm)	Log Reduction (24 h contact)	
			S. aureus	**K.pneumoniae**
None	0	0	-0.8	-1.3
	10	0	-1.5	-0.6
	20	0	-1.2	-0.8
	50	0	-1.2	-1.0
SILVADUR™ ET	0	256	1.0	1.6
	10	199	3.4	3.4
	20	169	4.1	>4.4
	50	129	3.8	>4.4

As is clearly evident from the results in Table 7, the SILVADUR™ ET treatment provides for a very durable antibacterial finish. Even after 50 wash cycles there is 50% of the silver remaining on the fabric and we observe >3 log reduction of both gram positive (*S.aureus*) and gram negative (*K.pneumoniae*) organisms.

SILVADUR™ ET was also evaluated for performance on synthetic textiles. A 100% polyester fabric was treated with SILVADUR™ ET using the padding and drying process. The starting silver loading was 150 ppm. The fabric was subjected to 50 home launderings using a home washing machine. The antimicrobial performance of the samples was determined by using the ISO20743 method, which is now the method of choice globally for silver containing textiles. The data from this experiment is shown in Table 8.

Table 8. Durability of SILVADUR™ ET on 100% polyester knit fabric

Finish	Wash cycles	ISO 20743 Log Reduction (24 hr Contact)	
		S. aureus	**K. pneumoniae**
None	0	-1.3	-1.8
SILVADUR™ ET	0	3.0, >3.8	>3.5, >3.5
	50	2.5, 3.5	3.0, 2.5

It is evident from examination of the results in Table 8 that the SILVADUR™ ET treatment provides for a very durable antibacterial finish on 100% polyester knit fabric. Even after 50 wash cycles we observe >3 log reduction of both gram positive (*S.aureus*) and gram negative (*K.pneumoniae*) organisms.

We have also treated a 100% PET/nylon (50/50) fabric with SILVADUR™ ET. In this experiment a wash durable package was included to dramatically improve the wash durability of the antibacterial finish. The wash durable additive is composed of small dosages of chemicals commonly used in textile processing. Samples of 100% PET/Nylon (50/50) were treated with SILVADUR™ ET using the padding and drying operation. The samples were then subjected to a number of home laundering cycles using a Launder-Ometer® and the AATCC method 61 type No. 2A. After the washing cycles the samples were analyzed for silver content and antibacterial efficacy measured by the AATCC Method 100. The results are tabulated in Table 9.

Table 9. Durability of SILVADUR™ ET on 100% PET/Nylon (50/50)

Treatment	Wash Cycles	Silver (ppm)	Log Reduction (24 h contact)	
			S. aureus	*K.pneumoniae*
None	0	0	-2.3	-2.7
	10	0	-1.6	-3.3
	20	0	-1.3	-3.0
	50	0	-1.0	-4.0
SILVADUR™ ET	0	389	2.7	>3.7
	10	324	>3.8	>3.4
	20	309	>3.7	>4.0
	50	245	3.6	4.0

As shown upon review of the results in Table 9, the SILVADUR™ ET treatment provides for a very durable antibacterial finish on a polyester/nylon fabric. Even after 50 wash cycles 60% of the silver remains on the fabric and >3 log reduction of both gram positive (*S.aureus*) and gram negative (*K.pneumoniae*) organisms is observed.

Exhaustion application

SILVADUR™ ET can also be readily applied to fabrics via the exhaustion process. We have found that by adjusting the pH of the treatment bath to ~5 a very high (>90%) efficiency of transfer of silver from the water to the fabric can be achieved. The amount of silver transferred to fabric via exhaustion under these conditions can be found in Table 10.

Table 10. Incorporation of SILVADUR™ ET via the exhaustion process

Fabric type	Liquor:fabric ratio	Concentration of silver in bath		Silver on fabric (ppm)
		Before (ppm)	After (ppm)	
Cotton	20:1	15	0.5	300
PET	33:1	12	0.6	360
PET	20:1	15	0.7	289

These results demonstrated that >90% transfer from the treatment bath to the fabric can be routinely achieved by carrying out the exhaustion at 90°C.

SILVADUR™ ET kills Methicillin-resistant Staphylococcus aureus (MRSA)

Methicillin-resistant *Staphylococcus aureus* (MRSA) is a <u>bacterium</u> responsible for several difficult-to-treat <u>infections</u> in humans. It may also be called multidrug-resistant *Staphylococcus aureus* or oxacillin-resistant *Staphylococcus aureus* (ORSA). MRSA is, by definition, any strain of <u>*Staphylococcus aureus*</u> bacteria that has developed <u>resistance</u> to <u>beta-lactam antibiotics</u>, which include the <u>penicillins</u> (<u>methicillin</u>, <u>dicloxacillin</u>, <u>nafcillin</u>, <u>oxacillin</u>, etc.) and the <u>cephalosporins</u>. MRSA is especially troublesome in hospitals, where patients with open wounds, invasive devices and weakened <u>immune systems</u> are at greater risk of <u>infection</u> than the general public.

Cotton fabric treated with SILVADUR™ ET at 100 and 300 ppm of silver was tested against MRSA.

Table 11. Efficacy of SILVADUR™ ET treated 100% cotton twill against MRSA

Sample	CFU/sample (0 h)	CFU/sample (24 h)	Percent Reduction of MRSA
Untreated	1.62×10^5	1.73×10^5	No reduction
Control	1.62×10^5	1.72×10^5	No reduction
0.3% SILVADUR™ ET	5.00×10^4	4.50×10^1	99.97
(100 ppm silver)	4.80×10^4	1.00×10^1	99.99
0.9% SILVADUR™ ET	4.00×10^4	1.00×10^1	>99.99
(300 ppm silver)	4.25×10^4	1.00×10^1	>99.99

The data in Table 11 indicates that even at 100 ppm silver the fabric sample reduced >99.9% of the MRSA organisms. Thus SILVADUR™ ET treated textiles are very effective in killing the MRSA organism.

Conclusion

The Dow Chemical Company has developed SILVADUR™ ET, a silver based product which provides a durable, light and heat stable antimicrobial performance to a wide range of textiles and nonwoven materials. The product utilizes polymer-Ag$^+$ complexation technology to deliver silver ions via a "smart control" mechanism thus providing for a light stable durable antimicrobial finish. Since the product is a liquid and readily water dilutable, it forms phase stable and solids free finishing bath solutions for easy and rapid processing of fabric. The material can be applied by both pad and exhaustion processes. Extensive testing of a range of treated materials has shown that the finish is very durable and provides antibacterial activity even after 50 wash cycles. The data presented in this paper suggests that SILVADUR™ ET is compatible with both natural and synthetic fabrics and provides a durable antimicrobial treatment without causing any detrimental effect on the feel of the fabric.

REFERENCES

[i] International Maritime Organization (2001). *Adoption of the Final Act of the Conference and any Instruments, Recommendations and Resolutions Resulting From the Work of the Conference. Final Action of the International Conference on the Control of Harmful Anti-Fouling Systems for Ships, 2001.* October 2001

[ii] Von Naegelli, V. *Deut. Schr. Naturfosch. Ges.* **1893**, *33*, 174-182

[iii iii] Foegeding, P. M.; Busta, F. F. in *Disinfection, Sterilizing and Preservation* (Block, S. S. ed), 4th Ed. **1991**, pp 803-832, Lea and Febiger, Philadelphia

[iv] Shimono, F.; Yamamoto, K.; Onishi,T.; Mioshi, R.; US Patent 5290544 **(1994)**

[v] Tanimoto, T.; Watanabbe, N.; Nakasima, K.; Matsuo, R.; Nagata, M.; Shingai, Y.; Otani, T. US Patent 6071542 **(2000)**

[vi] Ohsumi, S.; Kato, H. US Patent 5405644 **(1995)**

[vii] (a) Nomiya, K. *et. al. J. Inorg. Biochem.* **1995**, *58*, 259-267 (b) Nomiya, K. *et. al. J. Inorg. Biochem.* **1997**, *68*, 39-44 (c) Nomiya, K. *et. al. J. Chem. Soc. Dalton Trans.* **1998**, 1653-1659 (d) Nomiya, K. Takahashi, S. Noguchi, R. *J. Chem. Soc. Dalton Trans.* **2000**, 4369-4373

[viii] Nomiya, K. et al. J. Inorg. Chem. Biochem. **1995**, 58, 259-267S

[ix] ASTM D882-02 – Standard Test Method for Tensile Properties of Thin Plastic Sheeting

THE CONTROL OF ODOUR IN SPORTS AND FASHION WEAR

Mike Sweet
Polygiene AB
Stadiongatan 65, SE-217 62 Malmö, Sweden
Tel: + 46 40 530 202 Fax: + 46 40 530 210 email: info@polygiene.com

BIOGRAPHICAL NOTE

Mike Sweet, aged 57 has spent all his working life involved in textiles (dyeing and finishing) both in mill production capacities and as technical sales support. He started his career in Nottingham UK in the laboratories of a very large local dyehouse gaining experience of all aspects of textile `wet` processes.

Mike was educated at Nottingham Trent University and joined a German textile auxiliary manufacturer as technical sales, staying for 25 years and leaving as MD of UK operations.

Mike joined Polygiene (Sweden) in 2008 as technical R&D for textiles.

ABSTRACT

This paper will attempt to outline the current upturn in demand for odour free textile finishes.

Our lives are changing, there is now little if any discrimination between work clothes, sports clothes or weekend clothes, the boundaries are vague. Combine this with our ever busy lifestyles, and the demand for personal hygiene, and the need for help in maintain our clothing (ourselves) fresh and confident is evident. An additional argumentation may also be the need to wash articles less often, and therefore save energy and the environment, but this argument is often over stated. Reducing a wash temperature from 40c to 30c does not save significant amounts of energy.

Slide 1

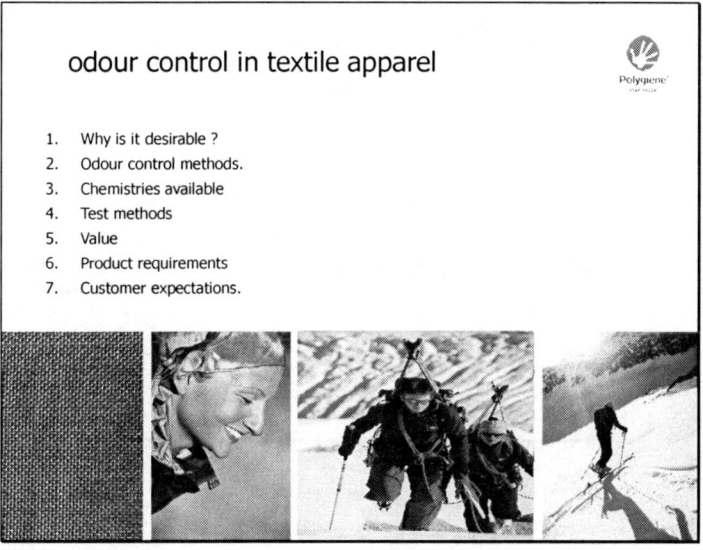

Slide 2

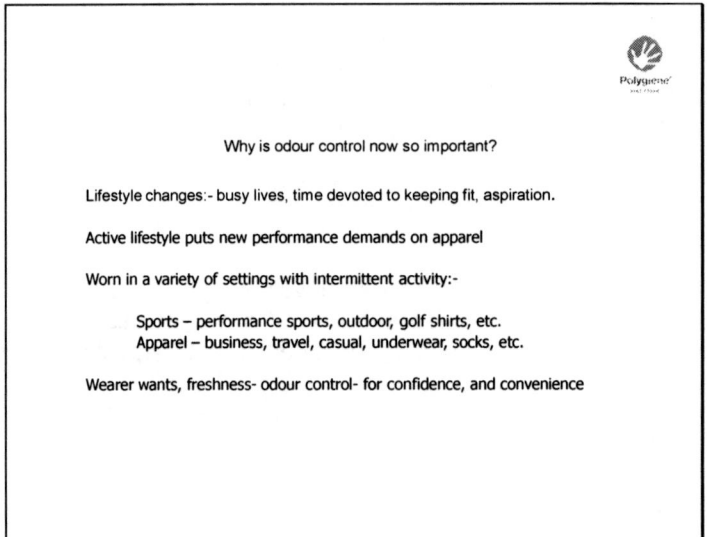

Market forces. Apparel manufacturers are constantly seeking USP or added value. The customer now is also value savvie and will actively seek out articles that seem to offer more value for money. In 2006 the global apparel market was estimated at £1,253 billion, with sporting goods alone valued at €400 billion estimated growth rate of 5% per annum that gives us almost £1,600 billion in 2010. Sales value at retail for global apparel (this is all apparel fashion, sports, formal etc) is estimated at $1,800 billion for 2010. Over 65% of these articles are now manufactured in China, Asia as a whole accounting for over 80% , but even so supply cannot keep up with demand.

Slide 3

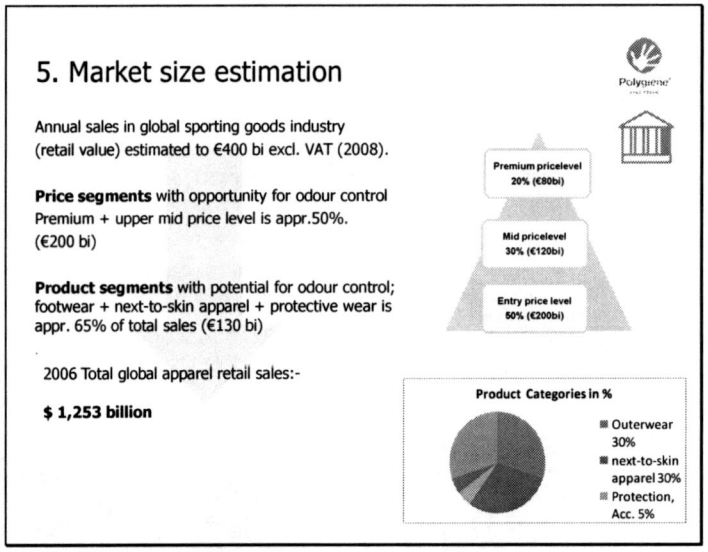

Some export values

Country	$ billions
China	8,261
Hong Kong	1,725
Italy	1,353
Germany	670
USA	600

It is plain to see that Europe and USA are now very much minor players

Slide 4

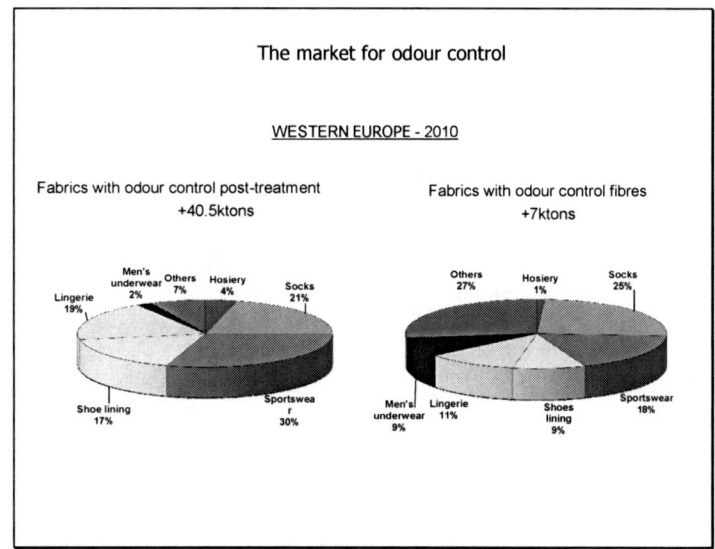

Pie charts showing break down of textile segment for odour control sales. Most popular area is sports wear (base layers ,running vest etc, etc). Most popular application method by far is topical treatment. Topical treatment does not tie the producer down to any one fibre supplier, yarn type or construction. Existing specifications can be simply upgraded to odour free with no change to any aesthetics of the article. Nor does it extend the supply chain or warrant extra stock holding.

Some of the fiber treatments produce yarns that are expensive, more difficult to weave or knit and are self coloured so are visually apparent in the finished article.

Slide 5

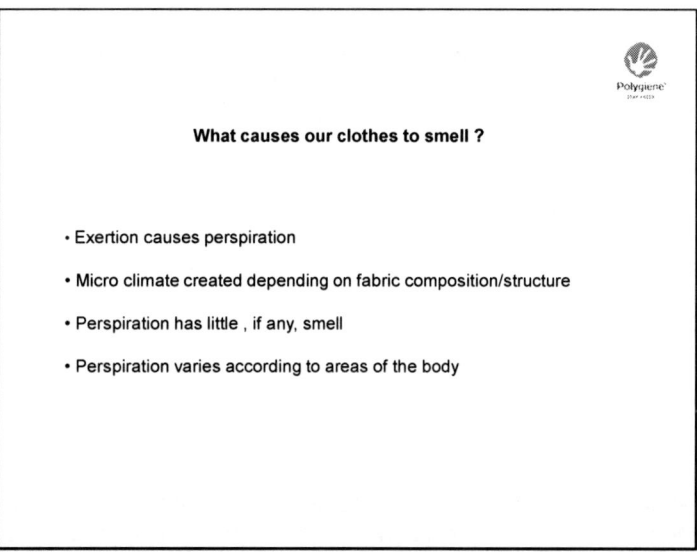

A healthy persons fresh perspiration does not normally smell. Perspiration performs several functions cooling and lubrication,as well as pheromone release.

Slide 6

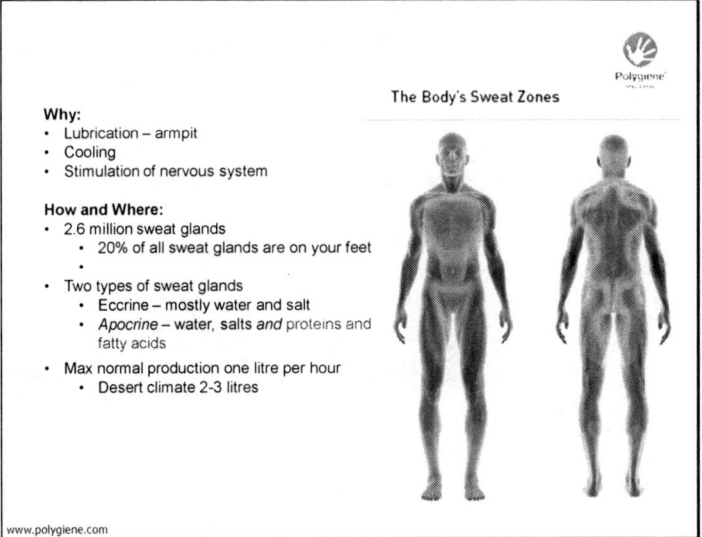

Schematic of sweat zones are shown in slide 6. Red zones produce the highest volumes. Chemical composition varies according to the zone of the body.

Slide 7

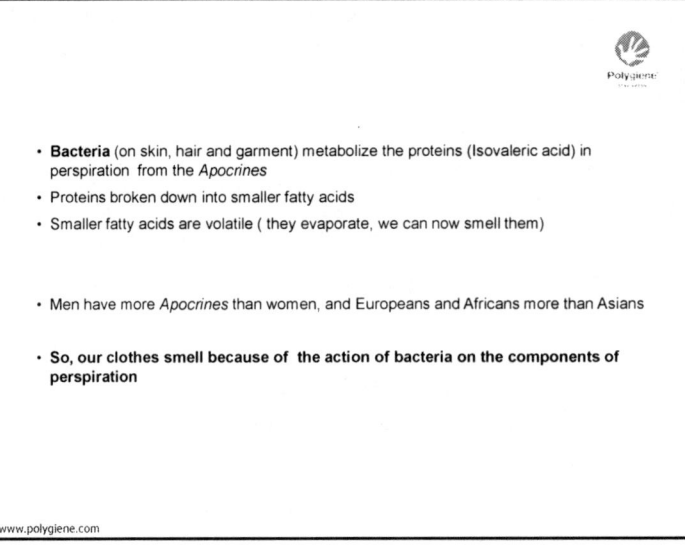

Our sweat is transferred to our clothing. Bacteria present breakdown the fatty acids and proteins to produce volatile compounds that we can detect. The level of odour can be related to the construction of the garment in terms of density, fibre type, chemical finish, body temperature, relative humidity, etc.

Slide 8

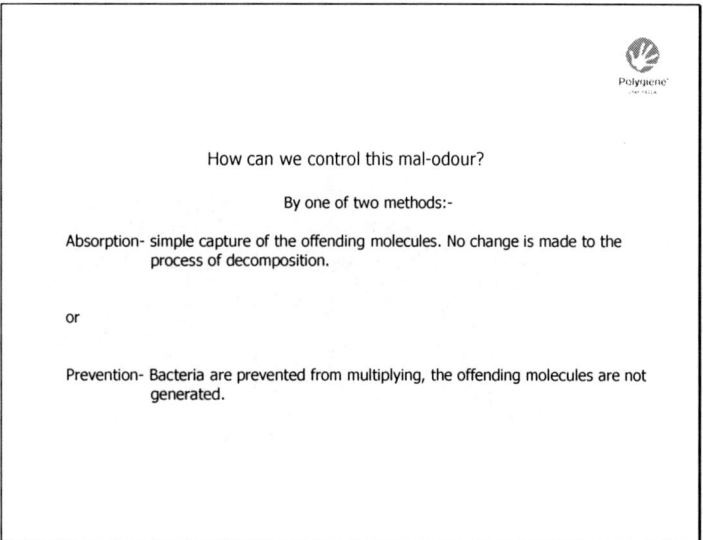

Two distinct control methods, i) simple adsorption or trapping ii) antimicrobial prevention. With absortion no chemical change occurs.

Slide 9

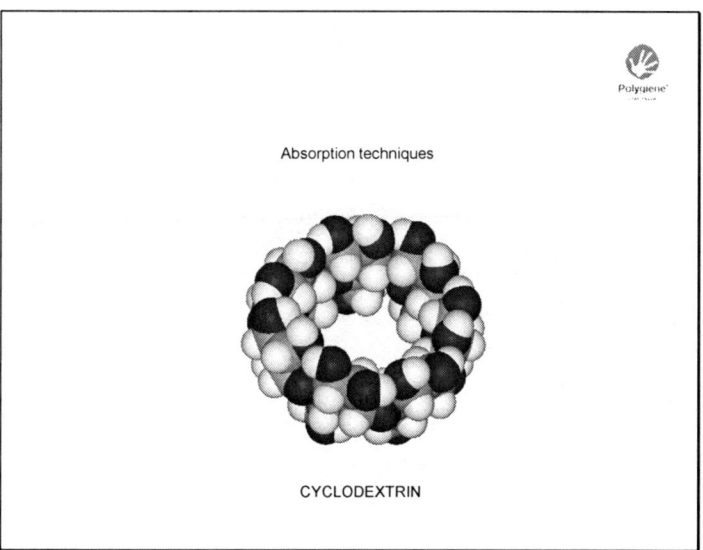

A schematic of a cyclodextrin molecule. The offensive odour molecule is physically attracted to and trapped with the centre of the ring.

A microphotograph of activated carbon (slide 10). Mal odours are trapped on the enormous surface area.

Both these physical approaches suffer several drawbacks:-

They are non discriminatory- they absorb any volatile, and they are active immediately they are exposed to any atmosphere, not just when the consumer first wears the garment.

In addition once `full` they cease to function. It is claimed they are emptied in the wash cycle but absolute cleaning is unlikely and performance falls off dramatically with repeated washing

Slide 10

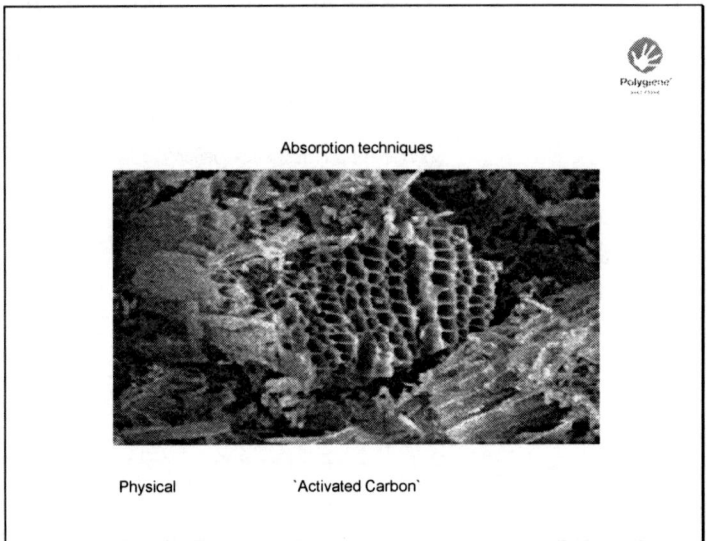

Some of the chemistries available for odour control.. Not all work on all fibres, some have aesthetic drawbacks, others are better antifungal than antibacterial. Knowledge of process conditions and customer expectations are required.

Slide 11

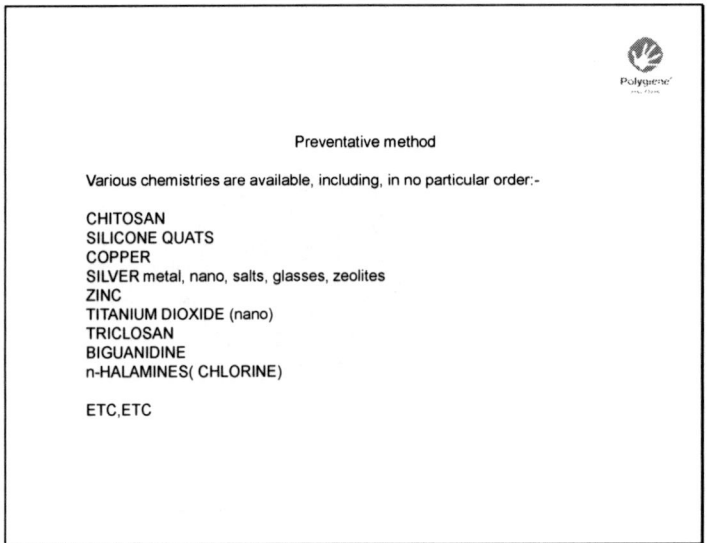

Fibre included antimicrobial.Not exhaustive. Antimcrobial is added at the fibre extrusion stage. Normally higher levels are required than found in a topical treatment.

Slide 12

Examples of antimicrobial yarns, past and present	
RHOVILAS ®	RHOVYL
AMICOR ®	COURTAULDS
AMICOR PLUS ®	COURTAULDS
SILFRESH ®	NOVACETA
MICROSAFE AM ®	HOECHST-CELANESE
BACTEKILLER ®	KANEBO
LIVERFRESH N ®	KANEBO
LIVERFRESH A ®	KANEBO
LUFNEN VA ®	KANEBO
SA 30 ®	KURARAY
BOLFUR ®	UNITIKA
FV 4503 ®	AZOTA-LENZING
CHITOPOLY ®	FUJI-SPINNING
THUNDERON ®	NIHO SANMO DYEING
X-STATIC	NOBLE

Application methods for topical treatments

Padding simple positive mechanical application. Textile is immersed in a solution of product of known concentration of product, squeezed between two rollers-reduces water content which lowers cost of drying and also forces product into the construction. Normally other finishing agents such as softeners/stitch lubricants-required to reduce needle damage when sewing- can be applied simultaneously. Folowed by standard drying. Padding is best suited to larger quantities of fabric. Not suitable for knitted fabrics required in circular form.

Slide 13

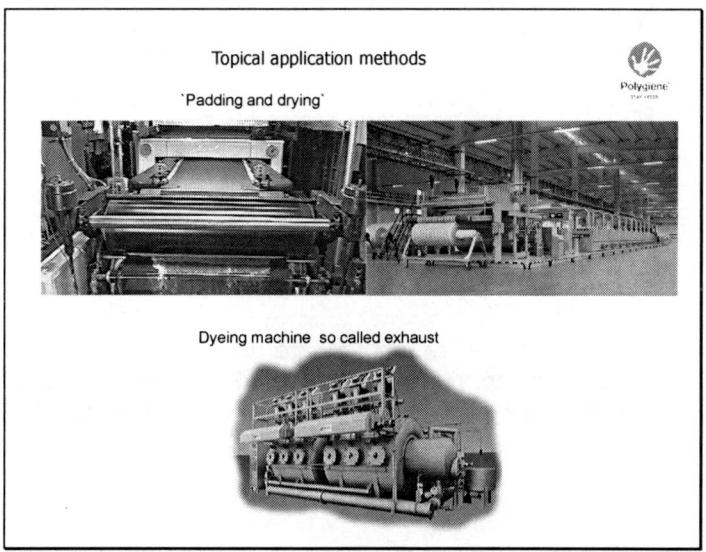

Exhaust application is when the product is applied as the final stage in a dyeing machine. Not all chemistries are suitable for this technique. Best suited to smaller batch sizes, less wastage. Exhaust application is a not a positive technique, by which I mean 100% application cannot be guaranteed and some product will be discharged when the machine is drained.

Slide 14

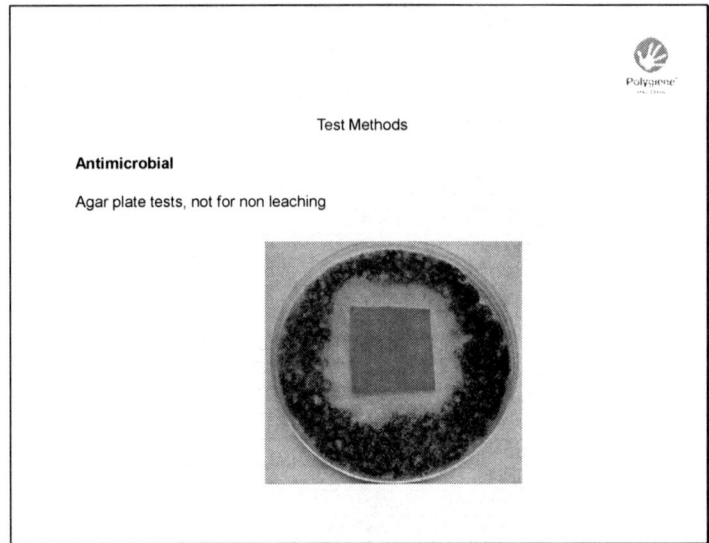

The wrong test method. Apparel now requires non migrating technology. It was a requirement for textiles to show a Zone of Inhibition, this is no longer desirable as shows possible migration to the skin.

Slide 15

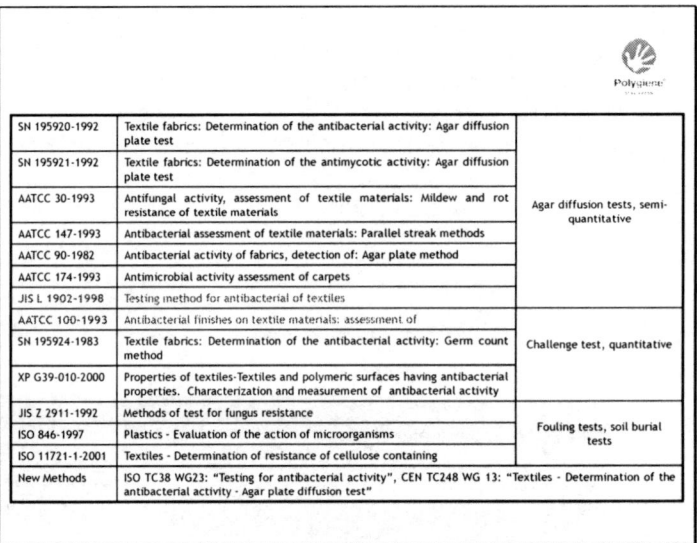

Some of the test methods available for textiles, not all are relevant to odour control techniques, highlighted are AATCC 100 and JIS 1902 which are the common methods. Others are available including the E-2149 dynamic shake. However it is actual field testing that is important rather than any artificial laboratory simulation.

Slide 16

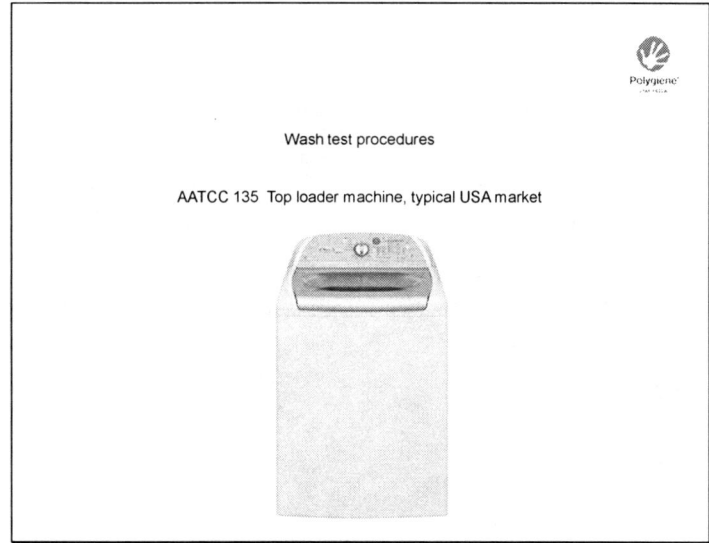

Wash test procedures for durability of finish. American standard is AATCC 135 which uses a typical USA top loading washing machine. Standard domestic wash temp is 40c. A lot of these older machines have no thermostat or heater fitted.

Slide 17

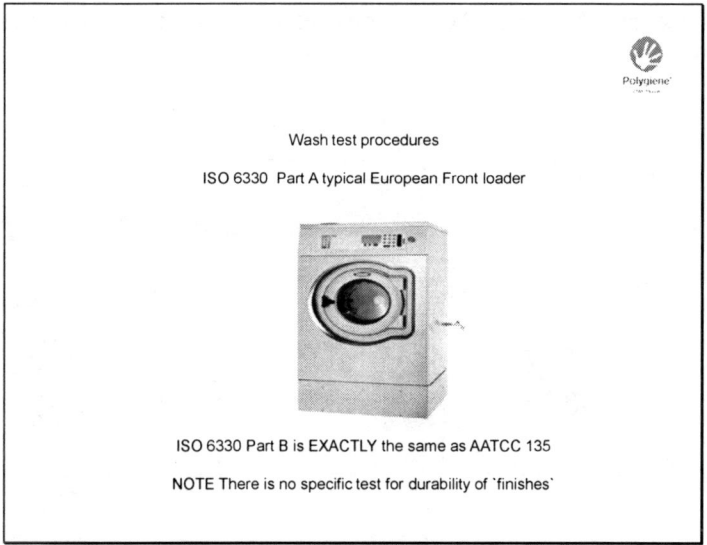

Typically Europeans would use a much smaller and more efficient front loader. Illustration is actually a commercial `Wascator` with computer programmed control of water level,temperature, agitation etc etc. European brands tend to request ISO 6330 Part B of this standard using a top loading machine is actually AATCC135. Note neither of these standards was actually designed to measure the durability of a `finish` to washing, they are specifically written for assessment of shrinkage, colour loss and appearance after multiple laundering.

Slide 18

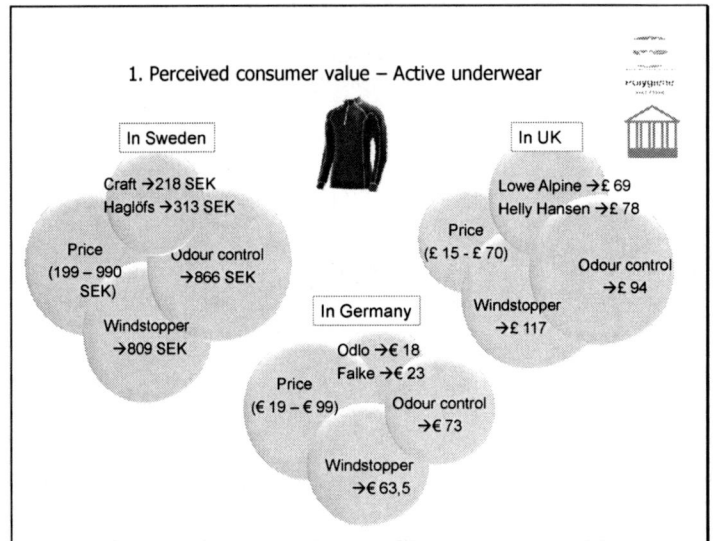

Results of straw polls outside suitable stores. Customers asked what price they would be prepared to pay a) for a brand b) for a performance brand c) for odour control. Results show that although customers are brand loyal, they expect the brand to deliver additional functions, and odour control is perceived as a major function. NB not many, if any of the customers realized that addition of odour control to textile articles was possible. They understood they could buy odourfree socks, but not other items. Also note it is mainly women who buy odour free mens socks, now what does that tell us? It is unlikely that this means brands can increase their retails price, simply saying they will pay more is not the same as actually doing it, but it shows that value is perceived.

Slide 19

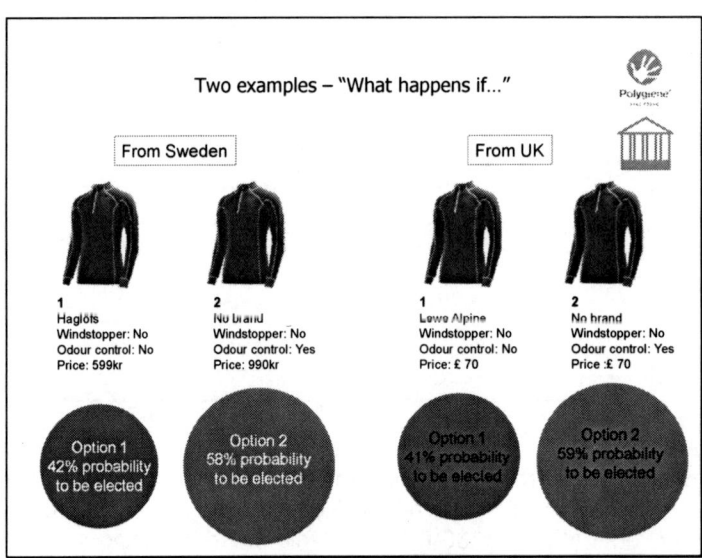

Confirmation of slide 18. In this case customers were simply asked to indicate their preferred garment 'specification' which they would like to buy regardless of the price. Most customers were prepared to pay more for added functionality. Odour control in this market is seen as a garment function, not as well being. The USA hunting/shooting market views odour control somewhat differently, to them it is another form of camouflage which supports the visual aspects of their clothing.

Product profile for an antimicrobial for application to apparel. It must be acceptable to the end customer-skin,environment To the brand price performance, and to the mill application methods. The supplier of the chemical must be able to show detailed support in all areas.

Slide 20

Suitability for textiles

The textile industry has a huge range of antimicrobial products to choose from but for odour control on apparel several important criteria must be met:-

NON MIGRATING

SKIN SAFE

MINIMAL ENVIRONMENTAL IMPACT, BOTH IN APPLICATION AND USE

REGISTRATIONS: BPD, EPA, OEKOTEX and BLUESIGN

PERFORMANCE- IN USE AND WASH DURABLE

EASE OF APPLICATION / COST

LITTLE IF ANY EFFECT ON OTHER PROPERTIES e,g WICKING, F/R, DWR etc

Various statements from trade and public press:-

- Consumers **expect** clothes that inhibit odour, just as they now expect garments that wick moisture"
  M Woolf, "Outdoor Summer 2006"
- "**Odour control** second most desired characteristic in performance apparel, just after fit and comfort"
  Market research
- "**Odour** has a fundamental influence on people's general **comfort** and even on their **health**"
  K Gniotek, "Odour Measurement in Textile Industry"
- Consumers more tuned to **hygienic life** style: "The consumers are now increasingly aware of the hygienic life style and there is a necessity and expectation for a wide range of textile products finished with antimicrobial products"
  Dr T Ramachandran, "Antimicrobial textiles – an Overview"

Slide 21

Odour control – the second most sought after feature by consumers

- "Consumers **expect** clothes that inhibit odour, just as they now expect garments that wick moisture"
  M Woolf, "Outdoor Summer 2006"

- "**Odour control** second most desired characteristic in performance apparel, just after fit and comfort"
  Market research

- "**Odour** has a fundamental influence on people's general **comfort** and even on their **health**"
  K Gniotek, "Odour Measurement in Textile Industry"

- Consumers more tuned to **hygienic life** style: "The consumers are now increasingly aware of the hygienic life style and there is a necessity and expectation for a wide range of textile products finished with antimicrobial products"
  Dr T Ramachandran, "Antimicrobial textiles – an Overview"

www.polygiene.com

Brand support required when it comes to labeling of the finished article. Care must be taken not to cross the boundaries into pesticide, cosmetic or medical device requirements.We are offering odour control not elimination, or disinfection.

Slide 22

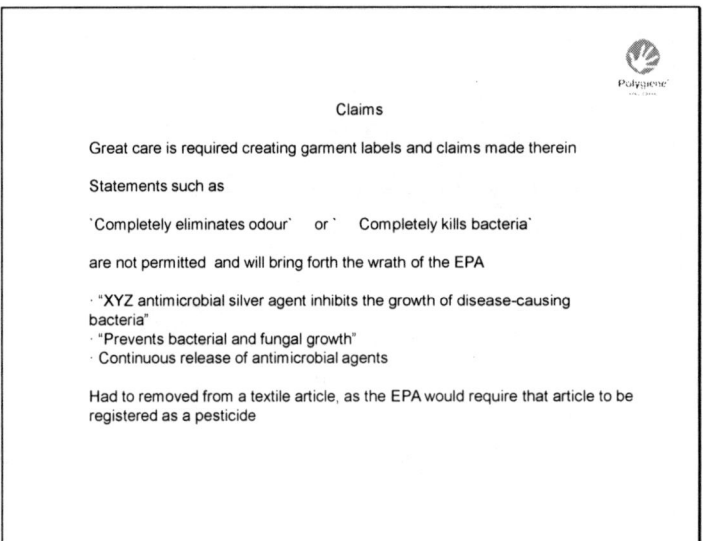

Slide 23

APPLICATIONS FOR SILVER IN TEXTILES AND MEDICAL DEVICES

Peter Steinrücke
BioGate
Neumeyerstrasse 48, Nuremberg 90411, Germany
Email: psteinru@web.de

+++ PAPER UNAVAILABLE AT TIME OF PRINT +++

ANTIMICROBIAL TECHNOLOGIES FOR CONTACT LENSES AND CONTACT LENS CASES

Manal M. Gabriel, D.D.S., Ph.D.
CIBA VISION Corporation
11460 Johns Creek Parkway, Duluth, GA, USA
Tel: +1 678-415-4289 Fax: +1 678-415-2092 email: manal.gabriel@cibavision.com

BIOGRAPHICAL NOTE

Dr. Manal M. Gabriel is currently Director, Innovation and Lens Care Research in Biomaterial Sciences department at CIBA VISION Corporation, where she has held various positions for the past eleven years. Manal has published many refereed articles, is an inventor on four patents, published five book chapters and numerous conference abstracts. Her research interests include: bacterial adhesion to medical devices, IOL and contact lenses; antimicrobial surfaces; deposit formation on contact lenses; acanthamoeba keratitis; and animal models for medical research purposes. Dr. Gabriel obtained her PhD degree in Microbiology and Genetics from Georgia State University in 1993 and DDS degree from Cairo University School of Dentistry. She is an adjunct professor at Georgia State University and is a member of ARVO, BCLA, ASM and TFOS.

ABSTRACT

Advances in contact lens and lens care technologies over the last few decades have further improved the safety and efficacy of contact lens wear. Even with the improved oxygen transmissibility of silicone hydrogel materials and the ease-of-use and more effective disinfection of today's lens care solutions, some adverse responses with contact lenses still occur. As such, the use of antimicrobial surfaces and materials for contact lenses, and lens cases, is being explored in an effort to further enhance the safety profile of contact lenses.

For decades, antimicrobial agents have been used with medical devices. A contact lens, or lens case with antimicrobial properties, could utilize a number of antimicrobial agents. Some of these could be applied to the surface of the lens, or to the surface of lens case material. Again, some of these could be incorporated into the bulk of the lens or into the bulk of the lens case polymer. Irrespective of the technology, the intended clinical benefit of an antimicrobial lens or lens case would be to reduce or eliminate adverse events related to infective agents; while simultaneously having minimal impact on the normal ocular flora.

The purpose of this presentation is to provide a brief overview of emerging antimicrobial technologies, some of which have potential for application with contact lenses and lens cases.

INTRODUCTION

When microbial contamination occurs, colonization is initially caused by bacterial adhesion to the lens or lens case surface.

Studies have shown a link between the incidents of Adverse Events (AEs) among contact lens wearers and contamination of their lens care systems.[1] McLauglin-Borlace study on bacterial biofilm in contact lens cases have shown that contamination of lens cases caused by bacteria, fungi and *Acanthamoeba* commonly occurs in 20-80% of CL wearers[2].

Numerous research and clinical cases have shown the association between bacterially contaminated contact lenses and the development of ocular complications, the most serious of which is microbial keratitis. In general, the source of bacterial contamination could be originated from the contact lens case, contact lens solution bottles, make-up cases (i.e. eyeliners and mascara), and use of water with contact lenses including swimming in lenses.[3,4,5]

Organisms isolated from contact lenses that are associated with corneal ulcers are often similar to those isolated from the patient's lens case but not in the same prevalence order[6]. The most commonly isolated ones are Gram negatives *Pseudomonas aeruginosa*, *Serratia marcescens*, and Gram positives *Staphylococcus* species which are less often isolated.

The spectrum of causative organisms in lens-related infections differs from that associated with non-lens-related infections, with up to 70% of culture-proven cases attributable to *P. aeruginosa* (opportunistic organisms), in contrast with non-contact lens wearers where *S. aureus* is the predominant bacterium. Willcox[7] provided a summary of the organisms isolated from corneal cultures reported by several authors.[8,9,10,11,12,13]

Table 1 shows the summary of the main bacteria isolated from MK with both HEMA-based lenses and silicone hydrogel lenses. The organisms appear similar for both lens types. Edwards, K. et al. 2004[14] reported the same types of bacteria isolated from AE.

Table 1. Summary of bacteria isolated from cases of MK with HEMA-based hydrogel and silicone hydrogel lenses.

Causative Bacteria	% Bacteria isolated from HEMA-based hydrogel lenses*	% Bacteria isolated from Silicone Hydrogel lenses
Total culture-positive cases	100	12
Gram Negatives	73%	75%
Pseudomonas aeruginosa	66%	42%
Serratia marcescens	4%	8%
Morganella morgani	1%	-
Escherichia coli	-	1%
Acinetobacter spp	-	8%
Alcaligens xylosidans	-	8%
Heamophilus influenzae	1%	-
Unidentified	1%	8%
Gram-positives	25%	25%
Streptococcus viridans	-	17 %
Coagulase-negative *Staphylococcus*	13%	-
6%		-
1%		-
3%		-
3% (mixed)		8%
1%		-

* The % bacteria isolated doesn't add up in all instances to 100 % because it is a compilation as reported by the authors of several different studies.

As reported in the literature, Table 2 provides information on bacteria commonly isolated from contaminated lens cases.

Table 2. Microorganisms commonly isolated from contaminated lens cases*

Type of Organism	Total organisms Isolated
Bacteria	
Staphylococcus epidermidis	29-59
Staphylococcus aureus	17-59
Streptococcus viridans	32-59
Klebsiella species	3-6
Pseudomonas aeruginosa	7-40
Enterobacter species	6-38
Acinetobacter calcoaceticus	2-3
Citrobacter amalonaticus	1-2
Bacillus subtilis	1-3
Other Gram negatives	18-58
Fungi and yeasts	
Cladosporium species	~41
Candida species	~38
Fusarium solani	Not given
Aspergillus versicolor	Not given
Exophiala spp	Not given
Phoma spp	Not given

*Gray, T. et al.1995 [6], Cabrera, J.V. et al. 1996 [15], Wilson, et al.1990 [16]

Acanthamoeba species are isolated from contaminated lens cases, gaining access to the case from the surrounding environment such as bathroom tap water. Contamination with *Acanthamoeba* have been isolated from roof storage water tanks in the United Kingdom, and India. [18] In the past few years, there have been two major outbreaks, *fusarium* keratitis and *Acanthamoeba* keratitis, which are rare but serious infections. Due to these outbreaks the FDA convened a meeting to reassess the guidance recommendations for testing lens care products before issuing FDA marketing clearance. The *Acanthamoeba* organism has not been a part of current panel organisms.

Possible Antimicrobial Technologies for Use in Contact Lenses or Contact Lens Cases

One potential method for preventing bacterial adhesion is the use of antimicrobial agents, which have been used in the medical field for decades. Antimicrobial agents have been used for a wide range of products including orthopedic implants, urinary catheters and wound dressings to name a few.[18-21] Currently, researchers are exploring the option of using antimicrobial surfaces and materials for contact lenses to further improve their safety. A wide variety of antimicrobial technologies could potentially be employed for use with a contact lens. Some may be applied to the surface of the lens material, while others may be infused directly into the lens polymer.

The technologies are aimed at preventing or reducing adhesion of bacteria to contact lenses or to contact lens cases as well as preventing biofilm formation by interfering with the bacteria signaling and communicating with each other.

The source of bacteria would originate from the environment, contaminated fingers, from contaminated lens disinfection solutions and/or contact lens cases. Controlling bacterial adhesion will help limit biofilm formation, potentially reducing the likelihood of adverse effects that are caused by bacteria.

An antimicrobial lens or lens case would need to be non-toxic to the human cornea and other tissue while providing broad-spectrum activity that kills or inhibits growth of undesirable organisms and reduces the colonization of the lens or case. Minimal impact on the normal ocular flora or activity of disinfection solution components would also be required.

Silver is currently used as the active agent on the surface of many antimicrobial medical devices. When used on the surface, silver slows the adherence and colonization of microorganisms by inhibiting DNA and RNA replication, disrupting the cell membrane, and interfering with cell respiration. It is very unlikely that specific bacterial strains could undergo simultaneous mutations to multiple mechanisms in order to develop resistance to silver's broad scope of action.

Silver is also currently being used in the contact lens industry. In fact, the FDA has approved a silver-impregnated contact lens case (CIBA VISION) for use with AQuify Multi-Purpose Solution.[21] When an aqueous solution comes in contact with the case, silver ions are slowly released, which provides antibacterial properties that will kill bacteria on contact. *In vitro* studies have demonstrated the case's efficacy against several strains of bacteria, including *Pseudomonas aeruginosa*. Clinical studies have demonstrated a significant forty percent reduction in the incidence of bacterial lens case contamination. [21]

Polymeric quaternary ammonium compounds (polyquats) are another option, they are surface-active agents. The efficacy in contact lens solutions is primarily caused by polyquats causing membrane damage allowing K^+ ions leakage from the cells. As a group they are considered antimicrobial and kill bacteria on contact. They are commonly used as disinfectants, preservatives, and algaecides for pools and hot tubs. Polyquats have also been used in contact lens solutions as disinfectants and preservatives. More recently, these compounds have been used in dental fillings, catheters, and polymers used in contact lenses to reduce bacterial biofilm formation and adherence to the surfaces of the devices.[22-24] In medical device applications and in contact lens solutions, use must balance toxicity with efficacy.

Polymeric pyridinium compounds can be covalently bonded to surfaces and have been shown to have a broad spectrum of antimicrobial activity. Upon contact with bacteria, the long amphipathic polycationic chains displace the divalent cations that hold together the negatively charged surface of the lipopolysaccharide network, thereby disrupting the outer membrane of Gram negative bacteria. Destroying the outer membrane permeability barrier, the cations further penetrate into the inner membrane producing holes and subsequently leakage of the cell contents. For Gram positive organisms, the cations have to penetrate the thick cell wall and reach the cytoplasmic membrane. In 2001, Tiller and colleagues were designing surfaces that kill bacteria on contact. The authors studied coatings made from two polymeric pyridinium compounds. They found that, in addition to having a broad spectrum of activity, the compounds did not leach from the material, so they were not depleted over time.[25] The percentage of bacteria killed on contact was higher than 99.8% for *Pseudomonas aeruginosa*. The toxicity for medical device applications is currently unknown.

Selenium compounds and nitric oxide-releasing polymers, free radical-producing agents, have been used for antimicrobial coatings as well. Selenium is a naturally occurring mineral, which can be found in many foods. It plays a key role in regulating the immune system. Selenium compounds can generate superoxide free radicals which can oxidize bacterial cells and prohibit cell growth. In 2006, Mathews and colleagues published a study investigating silicone hydrogel contact lenses with covalently bonded selenium in a rabbit model.[26] These lenses demonstrated resistance to *P. aeruginosa* colonization *in vitro*. Additionally, after 2 months of extended wear in rabbits, corneal health was not adversely affected by the selenium-coated silicone hydrogel contact lenses. Ozkan et al. conducted a 24 hour clinical trial to evaluate the antibacterial efficacy of Se-coated silicone hydrogel lenses *in vitro* and *ex vivo*, and assess cytotoxicity and clinical performance of Se-coated lenses. The results of the study suggest that Se-coated lenses are able to inhibit bacterial colonization. The overall clinical performance of the Se lenses was comparable to the commercially available lens, and the efficacy of Se-coated lenses is maintained after 24 hours of wear. [27]

Nitric oxide-releasing polymers also have antimicrobial properties, which are the result of oxidative and nitrosative stress caused by reactive intermediates of nitric oxide. In 2005, Nablo and colleagues studied nitric oxide-releasing coatings for stainless steel orthopedic implants, and found that these coatings can inhibit the adhesion of *P. aeruginosa*, *S. aureus*, and *S. epidermidis*.[19] Although the results are encouraging still NO release to the neighboring tissue remains a concern.

Quorum-sensing compounds are another class of agents with potential for use in antimicrobial coatings. The ability of microorganisms to communicate with each other and coordinate behavior is called quorum sensing. Subsequently, Quorum-sensing compounds inhibit bacteria by interfering with their signaling systems. Furanones (one example of quorum sensing compounds) are agents that occur naturally in red algae and prevent bacteria from colonizing on the algae's surface. The antimicrobial effect of adsorbed synthetic furanones on medical device polymers has been studied.[20] Baveja and colleagues have reported that a furanone-coated material significantly reduced *S. epidermidis* bacterial load on the polymer and slime production, while having no significant effect on the substrate's material characteristics. The use of furanones to coat contact lenses has also been studied. In one study, contact lenses were soaked in synthetic furanone, but the study results were unclear.[28] In another study, Zhu and colleagues found that fimbrolide used in coating contact lenses show promise as an antibacterial and anti-acanthamoebal coating and appear to be safe in an animal model.[29]

Anti-infectives, a different class of agents, also kill infectious organisms or prevent them from increasing in number and causing infection. Many antibacterial peptides are known to form pores in the lipid bilayers of microorganisms and cause a leakage of the organism's cell contents. In 2008, Willcox and colleagues demonstrated antimicrobial activity of contact lenses with cationic peptide coatings made of synthetic melamine and incorporating regions of protamine and melitin on the lenses.[30] They found that, when tested against *P. aeruginosa* and *S. aureus*, the coatings inhibited bacterial colonization by 70% for both bacteria. A 2007 patent by Ferreira et al. showed that the use of antimicrobial peptide (cysteine-incorporating cecropin-melitin hybrid peptide) immobilized on a planar surface exhibited antimicrobial properties after more than 3 weeks of storage in phosphate-buffered saline.[31]

The human body has potent anti-infectives that naturally occur from neutrophils and macrophages.[32] Defensins, small peptides that are rich in cysteine and one family of these naturally occurring anti-infectives can inhibit bacteria, fungi, and viruses. Defensins bind to the membranes of infecting organisms and increase permeability, decreasing the likelihood of resistance. Lactoferrin is another naturally occurring anti-infective. It is found throughout the body in mucous membrane secretions, such as saliva, tears, nasal and bronchial secretions, hepatic bile, and pancreatic fluids, and is essential for immune response. In a 2002 study, Singh and colleagues demonstrated that lactoferrin blocks *P. aeruginosa* biofilm development.[33] It is believed that lactoferrin acts by binding iron thus making it unavailable to the microbes as well as stimulating "twitching" of the bacteria, which prevents them from adhering to surfaces.

Discussion

Researchers in the contact lens industry have shown significant interest in agents that would provide antimicrobial properties for surfaces of contact lenses because they could reduce or eliminate the adherence of microbes to contact lenses and lens cases. Reducing exposure to infectious microorganisms could make contact lens wear possible for more patients and extended and continuous wear of contact lenses could improve convenience and increase acceptance of contact lenses as a vision care correction of choice. Patients could experience an added measure of protection from contamination without any extra effort on their part. An additional benefit is that bacterial resistance to many antimicrobial agents is unlikely because of their mechanisms of action.

Many aspects of antimicrobial technology as applied to contact lenses and lens cases must be considered in further research, such as whether antimicrobial lenses are compatible with lens care and whether antimicrobial agents could cause an allergic response. It is also unknown whether these agents would have unintended effects such as the build-up of endotoxins. Another concern is the cost of manufacturing antimicrobial lenses. These issues need to be adequately studied, and the answers will aid in the development of contact lenses that incorporate antimicrobial or anti-infective technology.

REFERENCES

1. Midelfart, J. et al., Microbial contamination of contact lens cases among medical students. The CLAO J. 1996; 22 (1): 21-24.

2. McLaughlin-Borlace, L. Bacterial biofilms in contact lens case. From web site: www. asmusa.org/edusrc/biofilms/infopage/056i.html. American Society for Microbiology.

3. Mayo, M.S., Schlitzer R.L., Ward, M.A., Wilson, L.A., and Ahearn D.G. Association of *Pseudomonas* and *Serratia* corneal ulcers with use of contaminated solutions. J Clin. Microbiol. 1987; 25(8): 1398-1400.

4. Wilson L.A., and Ahearn D.G. *Pseudomonas*–induced corneal ulcers associated with contaminated eye mascaras. Am. J. Ophthalmol. 1977; 84(1): 112-119.

5. Choo, J., Vuu, K., Bergenske, P., Burnham, K., Smythe, J., and Caroline, P. Bacterial populations on silicone hydrogel and hydrogel contact lenses after swimming in a chlorinated pool. *OVS*. 2005; 82 (2): 134-137.

6. Gray, T., et al. Acanthamoeba, bacterial, and fungal contamination of contact lens storage cases. British J. Opthalmology 1995; 79: 601-605.

7. M. Willcox, P. Sankaridurg, H. Zhu, E. Hume, N. Cole, T. Conibear, M. Glasson, N. Harmis and F. Stapleton. Inflammation and infection and the effects of the closed eye. *In:* Silicone Hydrogels Continuous-wear contact lenses, 2nd edition, edited by Deborah F. Sweeney, British Contact Lens Association, Butterworth Heinemann 2004; 90-125.

8. Weissman, B.A., Mondino, B.J., Pettit, T.H.,and Hofbauer, J.D. Corneal ulcers associated with extended – wear soft contact lenses . Am. J. Opthalmol. 1984; 97: 476-481.

9. Patrinely, J.R., Wilhelmus, K.R., Rubin, J.M., and Key, J.E. Bacterial keratitis associated with extended wear soft contact lenses. CLAO J.1985; 11: 234-236.

10. Mondino, B.J., Weissman, B.A., Farb, M.D. and Pettit, T.J. Corneal ulcers associated with daily wear and extended wear contact lenses. Am. J. Ophthalmol. 1986; 102: 58-65.

11. Cohen, E.J., Laibson, P.R., Arentsen, J.J. and Clemons, C.S. Corneal ulcers associated with cosmetic extended wear soft contact lenses. Ophthalmology. 1987; 94: 109-114.

12. Donnenfeld, E.D., Cohen, E.J., Arentsen, J.J. et al. Changing trends in contact lens associated corneal ulcers: an overview of 116 cases. CLAO J. 1986; 12: 145-149.

13. Schein, O.D., Ormerod, L.D. Barraquer, E. Et al., Microbiology of contact lens-related keratitis. *Cornea* 1989; 8: 281-285.

14. Edwards, K. Brian, G., Stretton, S., Stapleton, F, Willcox, M.D.P., Sankaridurg, P.R., Sweeney, D.F., Holden, B.A. Microbial keratitis and silicone hydrogel lenses. A look at the contributing factors, precautions and patient management issues with this complication. C L Spectrum. January, 2004; 38-43.

15. Cabrera, J.V. and Rodriquez, J.B. Ocular bacteria flora in contact lens wearers. *ICLC*. 1996; 23: 149-151

16. Wilson et al., Microbial contamination of contact lens storage cases and solutions. Am. J. Ophthalmol. 1990; 110:193-198.

17. Ahearn, D.G. and Gabriel, M.M. Contact lenses, Disinfectants, and *Acanthamoeba* keratitis. **In:** Advances in Applied Microbiol. 1997; 43:35-56. Academic Press.

18. Gabriel MM, Mayo MS, May LL, Ahearn DG. *In vitro* evaluation of the efficacy of a silver-coated catheter. Curr Microbiol. 1996;33:1-5.

19. Nablo BJ, Rothrock AR, Schoenfisch MH. Nitric oxide-releasing sol-gels as antibacterial coatings for orthopedic implants. Biomaterials. 2005;26: 917-924.

20. Baveja JK, Willcox MDP, Hume EBH, Kumar N, Odell R, Poole-Warren LA. Furanones as potential anti-bacterial coatings on biomaterials. Biomaterials. 2004;25:5003-5012.

21. Amos CF, George MD. Clinical and laboratory testing of a silver-impregnated lens case. Contact Lens Anterior Eye. 2006; 29(5):247-255.

22. Whiteford JA, Freeman WF. Methods and systems for preparing antimicrobial films and coatings. 2007. PCT WO2007/070801A2.

23. Majumdar P, Lee E, Patel N, Stafslien SJ, Daniels J, Thorson CJ, Chisholm BJ. Medical device coatings based on polysiloxanes containing tethered quaternary ammonium salts. Polymer Preprints (American Chemical Society, Division of Polymer Chemistry). 2008; 49(1):852-853.

24. Morris CA, Gabriel MM, Qui Y, Winterton LC, Lally JM, Ash MK, Carney FP. Medical devices having antimicrobial coatings thereon. US7402318B2

25. Tiller JC, Liao CJ, Lewis K, Klibanov AM. Designing surfaces that kill bacteria on contact. Proc Natl Acad Sci USA. 2001; 98(11):5981-5985.

26. Mathews SM, Spallholz JE, Grimson MJ, Dubielzig RR, Gray T, Reid TW. Prevention of bacterial colonization of contact lenses with covalently attached selenium and effects on the rabbit cornea. Cornea. 2006; 25(7):806-814.

27. **Ozkan J, Zhu H, Willcox W.** Efficacy and Clinical Performance of Selenium Antibacterial Silicone Hydrogel Contact Lenses. **Silicone hydrogels editorials. December 2009.**

28. George M, Pierce G, Gabriel MM, Morris C, Ahearn DG. Effects of quorum sensing molecules of *Pseudomonas aeruginosa* on organism growth, elastase B production, and primary adhesion to hydrogel contact lenses. Eye Contact Lens. 2005;31(2):54-61.

29. Zhu H, Kumar A, Ozkan J, et al. Fimbrolide-coated antimicrobial lenses: Their *in vitro* and *in vivo* effects. Optometry and Vision Sciences. 2008;85(5):292-300.

30. Willcox MDP, Hume EBH, Aliwarga Y, Kumar N, Cole N. A novel cationic-peptide coating for the prevention of microbial colonization on contact lenses. J Appl Microbiol. 2008; 105:1817-1825.

31. Ferreira L, Langer R, Loose CR, O'Shaughnessy WS, Zumbuehl A, Stephanopolous. Medical devices and coatings with non-leaching antimicrobial peptides. 2007. WO2007095393A2.

32. McDermott AM. The role of antimicrobial peptides at the ocular surface. Opthalmic Res 2008; 41:60-75.

33. Singh PK, Parsek MR, Greenberg EP, Welsh MJ. A component of innate immunity prevents bacterial biofilm development. Nature. 2002; 417:552-555.

CONSIDERATIONS IN THE DESIGN AND DEVELOPMENT OF AN ANTIMICROBIAL DELIVERY SYSTEM

Paul Lawrence & Debbie Stephenson
DuPont Teijin Films UK Ltd
P O Box 2002, Wilton, Middlesbrough, TS90 8JF, UK
Tel: 01642 572087   Fax: 01642 572083   email: Paul.D.Lawrence@GBR.dupont.com

BIOGRAPHICAL NOTE

Paul Lawrence is an Application Development Scientist in DuPont Teijin Films based in the Global R&T Centre, Wilton, UK. He has a strong background in Chemistry and Polymer Science and the key focus in his role today is to drive innovation by interacting directly with customers and their market applications. His core expertise is taking novel polyester films from conception and supporting their development through the entire supply chain from direct customers to end users. This is a very hands-on role which has required Paul to travel across the world, notably USA, Europe, and Asia Pacific, and in particular China. Career highlight must be the success of film developed and supplied to the 2nd Generation Identity Card Program for The People's Republic of China. This project won DuPont Teijin Films UK Limited a Queen's Award for Innovation and Paul was able to represent the company at a ceremony hosted by HM Queen Elizabeth II and held inside Buckingham Palace.

ABSTRACT

The successful design and market development of antimicrobial products (treated articles) is dependent primarily on having a sound understanding of the many factors affecting microbial survival on surfaces and the threats that such populations pose within a given application. But other factors come into play such as delivery system mechanisms, regulatory requirements, acquiring credible evidence for marketing claims and environmental influences. Many product designs and associated market claims are supported by data that has no relevance to the intended applications and in many situations the product simply cannot function. The data produced by DuPont Teijin Films with its Melinex® Shield polyester film development, is based largely on simulated use scenarios and realistic conditions of exposure have been used extensively. This has provided a source of information which has significantly affected the product design and its appropriateness in targeting harmful bacteria such as MRSA and EColi in healthcare, education, food processing and washroom environments. This paper reviews the many factors and considerations faced by DuPont Teijin Films in the design of its antimicrobial delivery system development.

SUMMARY

The successful design and market development of antimicrobial products (treated articles) is dependent primarily on having a sound understanding of the many factors affecting microbial survival on surfaces and the threats that such populations pose within a given application. But many other factors come into play such as delivery system mechanisms, regulatory requirements, acquiring credible evidence for marketing claims and environmental influences. Many product designs and associated market claims are supported by data that has no relevance to the intended applications and in many situations the product simply cannot function. The data produced by DuPont Teijin Films, with its Melinex® Shield polyester film development, is based largely on simulated use scenarios, and realistic conditions of exposure have been used extensively. This has provided a source of information which has significantly affected the product design and its appropriateness in targeting harmful bacteria such as MRSA and EColi in healthcare, education, food processing and washroom environments.   This paper reviews the many factors (see Figure 1) and considerations faced by DuPont Teijin Films in the design of its antimicrobial delivery system development.

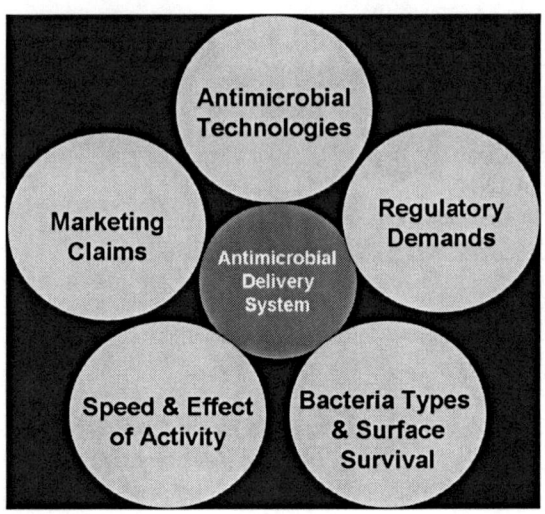

Figure 1: Factors influencing design of delivery system (DuPont Teijin Films)

ANTIMICROBIAL TECHNOLOGIES

Technologies in the marketplace

A wide range of technologies are available in the marketplace to achieve antibacterial effects in treated articles. For example, in waterborne paints, at least initially, antibacterial activity was achieved by increasing the concentration of additives used to provide in-can protection (eg chloromethylisothiazolinone or CMIT, and methylisothiazolinone or MIT) and some products manufactured in the Far East still employ this approach although the durability and functionality of the effect under normal conditions of use is generally considered to be very poor. Both organic (eg triclosan) and inorganic (eg silver nitrate) additives are used to generate antibacterial activity in treated articles. Probably the most common organic additives currently employed are; triclosan, pyrithiones, N-butyl-1,2 benzisothiazolin-3-one commonly known as butyl-BIT and 4,5-dichloro-2-n-octyl-4-isothiazoline-3-one or DCOIT. The most commonly employed inorganic additives are all based on the release of silver ions either through the use of salts entrapped in an inorganic matrix (eg TiO_2) or within a more complex structure (eg silver - zeolite complexes and silver in glass). Some limited use of copper ions exists mainly through the use of copper oxide. Copper, like silver, functions through the release of cations and will occur in a hydrated environment.

Choice of antimicrobial technology

The use of silver ion donors is by far the dominant technology in the market. Silver-based antimicrobial agents have been used by clinicians for over 100 years. They are considered safe and, at the concentrations used, have a very low environmental impact. For these reasons DuPont Teijin Films selected this technology for its development and selected its supplier based on commercial availability and regulatory approvals. To maintain the inherent very low environmental impact, DuPont Teijin Films has engineered its polyester film solution to release silver ions only when a situation in which a microbial hazard arises. Unlike some delivery mechanisms, the rate of release is regulated to accommodate the size and nature of the challenge presented. Silver ions do not possess a single mode of action but interact with a wide range of molecular processes within microorganisms resulting in a range of effects from inhibition of growth, loss of infectivity, to cell death. This mechanism depends on both the concentration of silver ions present and the sensitivity of the microbial species to silver. Contact time, temperature, pH and the presence of free water all impact on both the rate and extent of antimicrobial activity.

REGULATORY DEMANDS

The regulatory demands affecting antimicrobial products are clearly still work in progress. Very few countries have a framework in place to regulate the sale of treated articles. In the EU, the issues of articles with internal and external effects are very complicated. It has been agreed that if an article has an external effect, it should be regulated as a biocidal product, but how precisely to do this is not agreed. The preferred answer seems to be to regulate the active substance, or formulation used, and view the article as a delivery system.

The active substance used in the delivery system by DuPont Teijin Films is registered for use under the Biocidal Products Directive and is registered on the EU Provisional List of Additives used in Food Contact

Plastics. The carrier for this delivery system is based on one of DuPont Teijin Films existing base film products which already has regulatory EU and USA compliance for food contact articles. DuPont Teijin Films therefore set out to design the antimicrobial delivery system with sufficient concentration of silver ions present to work effectively and at the same time at levels so as not to compromise food contact regulations in order for the development to be suitable in food manufacturing, processing and preparation premises.

BACTERIA TYPES AND SURFACE SURVIVAL

As a polyester film manufacture, DuPont Teijin Films has found it necessary to understand the different types of bacteria and their complex mechanisms in order to design a delivery system that actually works. DuPont Teijin Films has proved the real need to have a sound understanding of which bacteria is the problem target, and what are the conditions it needs to survive, grow and pose a risk. Almost all bacteria are either gram negative or gram positive with differences being in the structure of the cell wall and the presence of an outer membrane in gram negative species.

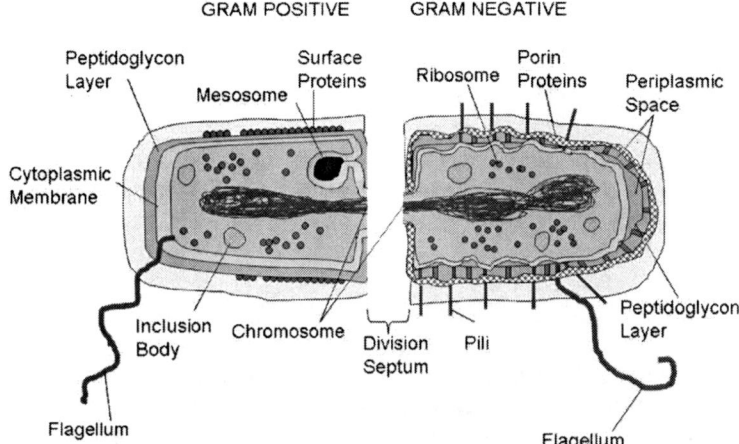

Figure 2: Bacteria Cell Structures

These cell structures will affect the uptake of antimicrobial ingredients such as silver ions and must be understood in order to determine the effectiveness of any delivery system in terms of lethal effect, speed of effect and longevity of effect. Microorganisms can be found on almost all environmental surfaces and, while most are harmless, others can cause disease in man and other animals and result in the spoilage of products that they come into contact with. Routine cleaning and disinfection are the main routes to remove these. However, some surfaces are either difficult to clean or are cleaned infrequently due for logistical reasons. Similarly, some surfaces are challenged too frequently for disinfection alone to be an adequate solution to the prevention of cross infection or microbial growth.

One of the key factors in producing a surface on which the maintenance of good hygiene can be facilitated is to ensure that it is smooth and lacks joints and crevices in which microbial populations can survive and, where conditions allow, grow. Engineering an antimicrobial agent into such interfaces provides an additional mechanism for the maintenance of good hygiene.

Most environmental surfaces in the clinical (domestic) environment are dry for the majority of the time. Although microorganisms might survive on such surfaces, they will not grow and many will die due to desiccation. Most such populations are small. However, there are occasions when a large microbial population does come into contact with a surface, for example from a splash of contaminated cleaning water, bodily fluid etc. When noticed, such splashes can be cleaned and disinfected but, when unnoticed, they can present a significant hazard under certain circumstances. Although these populations will probably decrease in size over time, the longer they remain viable, the longer the risk that cross-infection could occur remains.

SPEED & EFFECT OF ACTIVITY

Mechanism for antimicrobial effect

Once DuPont Teijin Films understood the basic cell structure and environments to target, it was important to understand the mechanism of the effect being delivered by the polyester film product in the development. In summary the rate of silver ion release is dependant on the contaminated drop size, the drying time and the soiling load as well as the bacteria species. There are three stages required:

1. the time taken for silver ions to be released from the film, reach the surface moisture and reach bacteria cells (being subject to competitive inhibition from any protein or other silver ion mopping-up agent present)

2. the uptake of silver ions and binding to cells
(Gram Negative more attractive to silver ions than Gram Positive but more silver is required to kill the former)

3. the transfer from binding of outer membrane to lethal effect inside cell
Gram Positive – silver ions penetrate cell wall, needing moisture and at least some metabolic activity to get into the cell; once in there, the bacteria cannot recover and die
Gram Negative – silver ions attach to outer membrane containing surface proteins and are not so dependant on moisture to attach to the membrane. However , the silver ions require moisture thereafter to transfer to and penetrate/damage cell wall causing leaking of cell in addition to intracellular effects.

Environmental Challenges

Microbial populations come into contact with inanimate surfaces in a number of ways either through the settlement of dust to which they are attached, through skin contact or following splashes of liquid contaminated with microorganisms (possibly bodily fluids in health care settings). In the latter situation, if the spill / splash are not cleaned and disinfected it will remain on the surface until it dries. This drying process will often result in a reduction in the number of viable bacteria carried but may leave a residue of viable propagules that can persist for extended periods of time (days to weeks in some cases; many years in others). After cleaning and disinfection many surfaces remain moist for extended period. Although residues of disinfectants, when they are present, can help prevent further microbial growth and low ambient temperatures can restrict the growth rates of many species, growth does occur. A number of environmental surfaces have been shown to be associated with the survival of bacterial populations and small reservoirs of species such as Listeria monocytogenes have been found in a number of food manufacturing units even after disinfection. The longer that a microbial population remains viable, the higher the risk that, should it present an infection hazard, cross-contamination / infection could occur. Similarly, in food production facilities, the presence of microbial populations can result from either contamination of food with a microbial population or the persistence of reservoirs of nuisance or hazardous organisms that become re-distributed during cleaning. It is possible that the inclusion of an antimicrobial agent into formulations used in such environments could increase both the rate and the extent of decline of a population presented in this manner and so both reduces the risk of cross contamination and prevent a residual viable population from remaining on the surface (or even proliferating).

Test Protocols

With all these factors now understood, the challenge for DuPont Teijin Films was to find a way to simulate the end-use performance and provide scientific evidence for marketing claims. The method most commonly used to explore potential antimicrobial activity is ISO 22196 (JIS Z 29801). It is a useful method for exploring potential antimicrobial activity but can also be readily modified to accommodate alternative microbial species and simulating alternative exposure temperatures. In some instances only minor modification would be required to produce a simulation of an end use to enable data to support a meaningful claim to be generated (eg the prevention of the growth of bacteria in the duct-work of air conditioning systems where condensation would be expected to form). But, in the majority of situations, both ISO 22196 and JIS Z 2801 as published, and even when modified to accommodate temperatures more appropriate to end-use, provide a poor simulation of most practical situations. However these methods can be modified further to simulate a number of scenarios that can occur in actual use and, under the guidance of certified independent microbiology laboratory, Industrial Microbiological Services Limited (IMSL), a set of new test scenario protocols were produced. Environmental challenges, as described in previous paragraph, can be simulated by a population being applied to the surface and then incubated at 65% relative humidity (RH). Samples removed at intervals can then be used to examine the survival / rate of kill over time. To replicate such scenario challenges, antibacterial activity on the delivery system from DuPont Teijin Films was determined using a method developed from JIS Z 2801: 2000 in which the impact of the treated system on a microbial population delivered as a splash of a contaminated liquid (or as a residue of a contaminated liquid) was studied. Both the exposure conditions (temperature, humidity etc) and the liquid employed were varied to simulate a range of environments (eg, clinical, food production, washroom etc) and multiple time points were used to examine the impact of the treatments on the rate of decline / growth of the populations.

An aliquot (typically 100µl) of a log phase cell suspension of either MRSA (ca 10^6 cells ml-1) or E coli (ca 10^6 cells ml-1) suspended in a sterile simulated splash material relevant to the use scenario (eg 1.5% bovine serum albumin - BSA to simulate bodily fluid or residue of a protein based food , sterile distilled water or sterile artificial urine solution) and inoculated onto replicate film samples and left uncovered for up to 24 hours at a temperature and relative humidity relevant to the end use scenario (eg 20°C and 65% RH). The size of the surviving population on replicate samples selected at random from those inoculated as above were determined at intervals (eg 1, 3, 6, 12 and 24 hours).

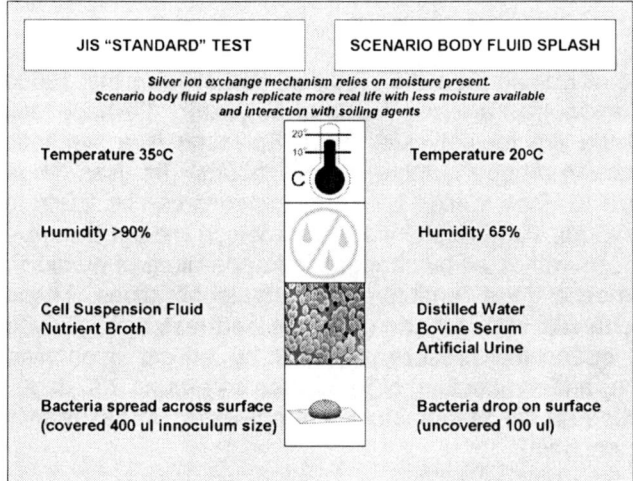

Figure 3: Comparison of Exposure Conditions

DuPont Teijin Films has set out to meet the environmental challenges previously mentioned and used test protocols simulating populations of bacteria reaching the surface as splashes of contaminated water, urine and sera.

MARKETING CLAIMS

It is perceived that the addition of antibacterial properties can be used to add value to systems employed where hygienic finishes are required. The addition of true benefit rarely accompanies this and where data is presented showing quantitative effects against bacterial populations, the data is often generated using the Standard JIS Z 2801 (ISO 22196) method. In very few instances is consideration given to the relevance of the test conditions to the end use. In some cases a 2 log10 reduction (99%) is considered to represent a "pass" level for antibacterial activity whereas in other instances reductions of 95% (from the starting population) are claimed to show "strong antibacterial activity". In addition, the actual ability of a population to survive on a substrate without the presence of an antimicrobial additive has not been examined and when investigated rigorously, in many cases no difference can be detected.

The efficacy of Melinex® Shield has been explored fully using tests that simulate the end use in which the product can be applied and can achieve 5 log10 reduction (99.999%). Testing was also carried out on polyester film without any additive present and this demonstrated no antimicrobial efficacy. This supports the claim it is the actual inclusion and design of the antimicrobial delivery system on the film surface that can prevent the survival and growth of harmful bacteria populations.

ANTIMICROBIAL DELIVERY SYSTEM

DuPont Teijin Films believes that polyester film (PETF) is an ideal robust carrier for an antimicrobial delivery system, and Melinex® Shield can be available with a pressure sensitive adhesive system for ease of covering flat surfaces. Such surfaces covered with Melinex® Shield can eliminate many of the crevices where microbial activity can grow and also provide a one-time cover for logistically difficult to clean areas.

Who is DuPont Teijin Films and what is polyester film?

DuPont Teijin Films is the world's leading supplier of polyester films with production plants all over the world and is a 50:50 joint venture between DuPont and Teijin Limited. DuPont is a science company. Founded in 1802, DuPont puts science to work by solving problems and creating solutions that make people's lives better, safer and easier. Operating in more than 70 countries, the company offers a wide range of products and services to markets including agriculture, nutrition, electronics, communications, safety and protection, home and construction, transportation and apparel. Teijin Limited is a major multinational enterprise offering fibres, chemicals and plastics, pharmaceuticals and medical products and diversified products. Teijin is using its proprietary technologies to expand into areas such as health-care products and services, advanced materials and information media.

Most people know polyester as a man-made fabric that transformed the clothing market from the 1960s onwards. Others might know that most clear plastic drink bottles are also made of polyester. Perhaps less obvious is just how many day to day applications there are for polyester film. Polyester is a synthetic polymer and, as a film, has been used for over 50 years in a huge number of applications. Its uses range from X-ray films to yoghurt lids, from diabetes test strips to video tapes. Latest innovations can be found in packaging, electronics and photovoltaic markets. Polyester film is essentially inert which means that is a perfect base for innovative surface coatings and is largely unaffected by changes in temperature of humidity. As it is lightweight and flexible, polyester film becomes a great "enabler" for many applications. These primary advantages such as high thermal stability, mechanical strength and chemical inertness are achieved by manufacturing a semi-crystalline film via a roll quenching process followed by biaxial orientation (stretching the film in machine and transverse directions) and heat setting at temperatures around 230 degC. See Figure 4 – Film Manufacturing Process. (Information on the manufacturing of polyester films is readily available on request via email to europe.films@gbr.dupont.com)

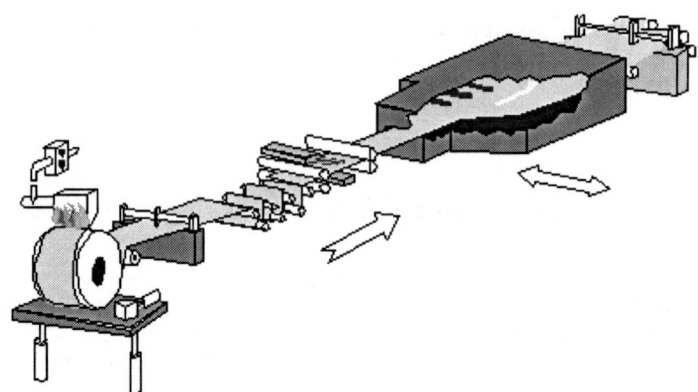

Figure 4: Film Manufacturing Process

Designing the antimicrobial delivery system

Before DuPont Teijin Films fully understood the dynamics of a delivery system, its first attempt to design an antimicrobial surface effect polyester film was to simply add the additive to a polymer matrix and process through standard conditions. This route maintained the transparent properties of the finished film, used sufficient raw materials to meet the necessary market cost model and was anticipated to provide an effective and durable functionality.

Film samples were produced and despatched to an external laboratory to carry out the standard JIS Z 2801 test protocol and they were returned with a positive 3 log10 reduction reduction after 24 hours for both MRSA and Ecoli (or >99.9% reduction from the starting populations of MRSA or Ecoli). Thinking this was a good indicator; DuPont Teijin Films proceeded to test the polymer route samples with the simulated splash test method as described earlier. After 24 hours the results for both bacteria species (MRSA and Ecoli) demonstrated a disappointingly ineffective 1 log10 reduction.

So began the journey for DuPont Teijin Films to fully understand the influencing factors and relationships between bacteria, technology, environment and "invisible" polyester film surface. The functional antimicrobial surface was characterised by DuPont Teijin Film's state of the art surface metrology suite. Surface design and particularly surface area coverage was quantified by employing white light interferometry and additional close-up morphology studies of the active species carried out by atomic force microscopy. This allowed the technical team to identify the key shortfall of the polymer design, ie the active silver particles were covered by polymer and not sufficiently close to the surface to be effective in the more demanding but realistic scenario environment.

The technical team needed to develop a design that could meet the following criteria:

- Ensure active silver particles could be freed from surface

- Ensure the silver particles did not fall off the surface

- Deposit sufficient silver particles to produce a long-lasting reservoir

- Be produced with our existing process capability

- Meet cost model for the marketplace

- Retain the desirable high gloss transparent appearance

- Meet current regulations

- Produce 3 log10 reduction (reduction of 99.9% from starting population) in simulated scenario testing

As an integrated polymer and film producer, DuPont Teijin Films (DTF) had the unique capability of being able to redesign the film effects through polymer recipe control and coating systems. The technical team tried coating the silver ions in a binder system onto the surface of the film but found the surface would be easily removed and compromised by even a gentle finger rub. So this coating route had to be rejected as unable to provide durable longevity or controlled migration functionality. The film was then re-designed by incorporating a hybrid coating/polymer mix and changing the crystallisation characteristics; this being achieved by pushing the boundaries of DTF process capability. Using the scientific techniques available to DuPont Teijin Films, the revised Hybrid design was compared with the original Polymer design and benchmarked against the criteria mentioned above.

Surface Effect

Surface electron microscopy images of the Hybrid film, showing exposed antibacterial particles and Polymer film which shows polymer covering over the active antibacterial particles.

Figure 5:

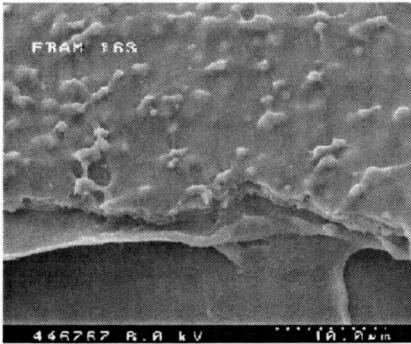

Polymer route

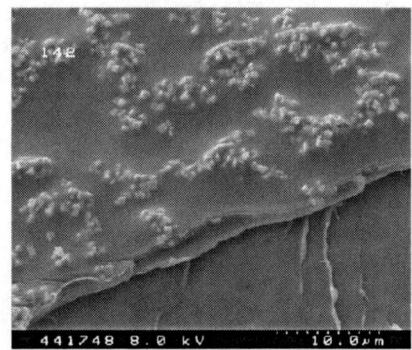

Hybrid route

Outcome – The Hybrid route successfully indicates a surface appearance showing the active silver particles are both free to be released but anchored until required.

Appearance and Cost

Based on ASTM D method, the average values of Polymer Film and Hybrid Film were taken for haze and total luminous transmission (TLT) and found to be very similar.

Figure 6

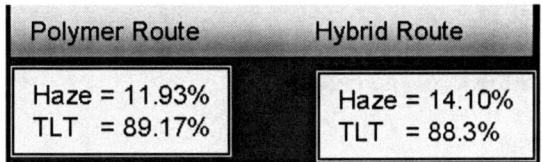

Polymer Route	Hybrid Route
Haze = 11.93% TLT = 89.17%	Haze = 14.10% TLT = 88.3%

Outcome –The Hybrid route was not detrimental to the aesthetic look of the film and allowed the film to retain a desirable high gloss transparent appearance. The haze value was also a reasonable measure for levels of additive present and with similar values this indicated equal quantities present and therefore no adverse effect on cost model from the hybrid design.

Regulations

DuPont Teijin Films works together as a leveraged competency with strong product stewardship and regulatory resources within its businesses, regions and corporation. It seeks a broad, diverse stakeholder base to enable better informed business decisions, viewing regulators as critical to business success. At each stage in this development, the technical team has regularly sought product stewardship and regulatory advice, both corporately and externally and is satisfied that Melinex® Shield meets existing EU biocide regulations today.

The carrier for this delivery system is based on one of DuPont Teijin Films existing base film products which already has regulatory EU and USA compliance for food contact articles and so DuPont Teijin Films commissioned an independent approved laboratory to test specific migration of silver in film samples using migration simulants and conditions as defined in EC Directive 97/48/EC,. The test required specific migration of silver using the single sided cell exposure method, into simulant B(3% acetic acid (w/v) in aqueous solution); exposure conditions of 1 hour at 40 degC. The level of silver in the extracts was determined using ICP-AES.

The limit specified in EC Directive 2002/72/EC for the migration of silver is 0.05 mg per kg of food simulant (ie 50 ppb).

The result detected from the DuPont Teijin Films sample determined the migration of silver to be <0.01 mg per kg of food simulant (ie <10 ppb)

Outcome – This is positive indication that DuPont Teijin Films has engineered its polyester film solution to provide a controlled release of silver ions and, in doing so, not exceed migratory levels specified in EC Directive 2002/72/EC. Unlike other plastic film solutions or migratory technologies, such a design combination of base film and delivery system provides DuPont Teijin Films with a platform to potentially demonstrate compositional compliance with food contact regulations in Europe and USA.

Simulated Scenario Tests

Plot of the logarithm of the number of remaining bacteria (Ecoli) in protein solution against time for the original Polymer and redesign Hybrid films – Simulated Splash Scenario Test Protocol

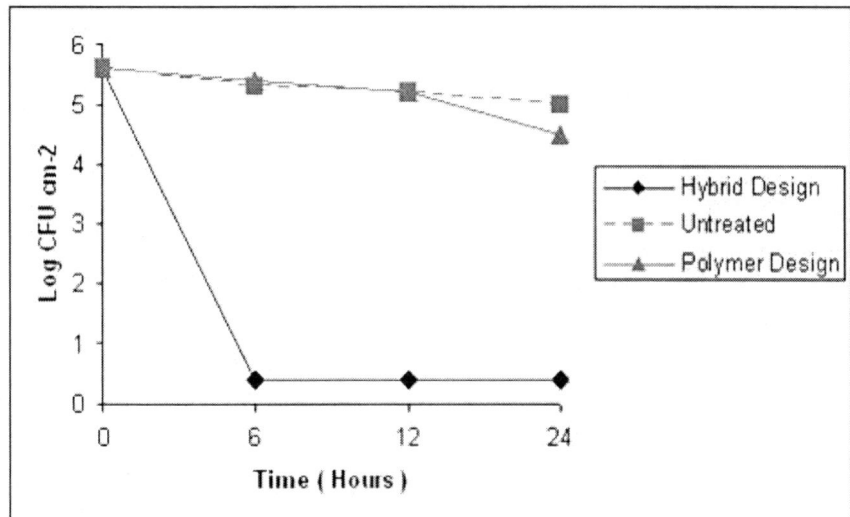

Figure 7

Outcome – The Hybrid route dramatically outperformed both the Polymer film design and untreated film design by achieving a 5 log10 reduction (99.999%) after 6 hours and successfully exceeded the target of 3 log10 reduction (>99.9%) after 24 hours. Similar results were seen for MRSA bacteria species and other simulated contaminated splashes (as described in Figure 3). This data supports the claim it is the actual inclusion and design of the antimicrobial delivery system on the film surface that can prevent the survival and growth of harmful bacteria populations.

CONCLUSION

Having a sound understanding of the factors affecting an antimicrobial delivery system is an essential requirement for a successful product design and for providing data that links claims, effects and benefits in a way that is vital both for regulatory approval and sustained credibility in the marketplace.

DuPont Teijin Films started with existing manufacturing process and very limited antimicrobial knowledge. It has been able to demonstrate its innovative approach into meeting real market challenges. This has led to a novel process capability able to provide active functional surface effects onto inert polyester film base. The development has also enabled DuPont Teijin Films to enter into a new market area for polyester film supported by marketing claims based on scientific evidence.

ACKNOWLEDGEMENTS

Julian Robinson, Research Scientist, DuPont Teijin Films
Peter Askew, Industrial Microbiological Services Limited, UK

Disclaimer

This information corresponds to our current knowledge on the subject. It is offered solely to provide possible suggestions for your own experimentations. It is not intended, however, to substitute for any testing you may need to conduct to determine for yourself the suitability of our products for your particular purposes. This information may be subject to revision as new knowledge and experience becomes available. Since we cannot anticipate all variations in actual end-use conditions, DuPont Teijin Films makes no warranties and assumes no liability in connection with any use of this information.

Nothing in this publication is to be considered as a license to operate under or a recommendation to infringe any patent right.

Caution: Do not use in medical applications involving permanent implantation in the human body.

For other medical applications, see "DuPont Teijin Films Medical Caution Statement", H-50102-DTF.

Melinex®, Kaladex®, Mylar®, Cronar® and Teonex® are registered trademarks of DuPont Teijin Films.

Only DuPont Teijin Films makes Melinex® brand, Kaladex® brand, Mylar® brand, Cronar® brand and Teonex® brand films.

FACTORS AFFECTING BIOCIDE COMPATIBILITY AND PERFORMANCE IN PVC DURING PROCESSING AND PRODUCT LIFETIME

Mr Dean Nichols
Akcros Chemicals Limited
P O Box 1, Lankro Way, Eccles, Manchester, M30 0BH, UK
Tel: 0161 785 1141 Fax: 0161 789 3186 email: dean.nichols@akcros.com

BIOGRAPHICAL NOTE

Dean Nichols has worked in the biocide industry since 1991 in technical, commercial and product development roles for major biocide manufacturing companies. Two years ago he joined Akcros Chemicals Ltd as their European Biocide Product Manager.

He oversees the development of new market areas and ensures that established biocides for plastics business continues to grow through a changing regulatory environment.

ABSTRACT

The use of biocides in PVC is commonplace but the choice of an appropriate biocide is not always easy. They are either added to protect against spoilage effects of microorganisms such as a reduction in aesthetic appearance or physical/chemical problems, or added to convey antimicrobial or hygienic properties to a PVC surface.

Biocides can be selected on the basis of their mode of action and the application for which they are intended but these aspects decision will not control the problems associated with microorganisms in every case.
This paper aims to highlight compatibility issues and establish the factors affecting biocide performance in PVC, both during processing and product lifetime. It will review a variety of active ingredients and experimental data generated by the UK laboratories of Akcros Chemicals.

Introduction

Biocides are added to treat a variety of synthetic materials including PVC. They are used traditionally to prevent degradation of the plastic itself or more recently to impart an external anti-microbial or hygienic effect. Biocides achieve this by controlling the survival or proliferation of microorganisms that come into contact with the PVC.

Where conditions for growth exist (appropriate temperature and levels of moisture) bacteria, yeast and mould may colonise and degrade the PVC by utilising raw materials within the plastic as a source of nutrients. Microbial growth can result in surface staining, pitting, reduction of structural strength, embrittlement, change in conductivity or flexibility and other physical or mechanical property changes. The growth of some microorganisms on PVC can cause the generation of malodours and the production of mycotoxins. Even where a microbial population is merely surviving, the potential for transfer to a more favourable substrate, such as food or a patient's wound, is undesirable.

In external environments algae can disfigure plastic surfaces. They do not utilise any of the PVC's components as a nutrient source, however algae harbours water which will promote physical 'freeze and thaw' effects and may permit growth of other microorganisms.

Therefore in many PVC applications the necessity for incorporation of a biocide is highly important.

The use of biocides in PVC

Applications where biocides are commonly used include: tarpaulin, canopies, pool and pond liners, silage pit liners, artificial leather, awnings, textile coatings, flooring, wallcoverings, signage, roofing materials, garden furniture, refrigerator gaskets, automobile components. Biocides for these uses are normally not intended to offer any hygienic effect but merely prevent the effects of microbial growth on or in the plastic itself.

The type of plasticiser used in the PVC formulation will influence the susceptibility of that PVC to microbial attack. Different organisms have varying abilities to utilise the plasticiser as a food source. It is known that phthalates are highly susceptible. Many organisms are of concern (Webb, 2000) and are of more numerous than those used in the laboratory test methods such as *Aspergillus, Chaetomium, Paecilomyces, Penicillium and Trichoderma* species.

Biocides added to confer hygienic effects to products such as toilet seats, kitchen utensils and medical devices do not necessarily provide any protection to the plastic article itself, although by their nature, could do so. Such applications provide activity against organisms that may come into contact or settle on them, thereby limiting potential transmission of disease-causing microorganisms.

For application areas such as hospital furniture, theatre flooring and coated textiles as well as perceived added value products such as sports clothing, bedding, chopping boards, worktops and other items for food preparation, the majority of biocides are targeted towards anti-bacterial applications. They are designed to enhance hygiene. Healthcare environments focus on pathogenic bacteria such as Methycillin resistant *Staphylococcus aureus* (MRSA), E.coli and other enteric organisms.

Measures of Biocide Performance

As a basic indication of microbiological performance the MIC (minimum inhibitory concentration) may be used to provide a rough comparison between biocide active ingredients. This can aid a preliminary choice of the most suitable active for a given application or against a specific organism. However, an MIC cannot be used in isolation. Other factors including bioavailability of an active ingredient, in-process stability and compatibility, ability to diffuse or migrate, leachability, UV and heat stability and biocidal performance should be considered.

Table 1: MIC (minimum inhibitory concentration) values of some biocide actives

Biocide active / micro-organism	Akcros NTERCIDE®	Alternaria alternata	Aspergillus niger	Trichoderma viride	Aureobasidium pullulans	Chaetomium globosum	Cladosporium cladosporoides	Sclerophoma pityphila	Penicillium glaucum	Pseudomonas aeruginosa	Staphylococcus aureus
OBPA	ABF	10	10	10	10	10	10	10	10	10	10
OIT	OBF	1.5	5	-	0.5	10	-	-	2.5	500	10
DCOIT	DBF	10	5	100	50	5	5	100	15	13	5
IPBC	IPBC	2	2	100	1	5	2	1	1	-	200
Zinc Pyrithione	ZNP	7.5	100	50	15	20	5	5	50	400	<10
Folpet	FLPT	-	100	500	75	100	-	-	50	1000	50
Triclosan	-	-	-	-	-	-	-	-	-	>100	0.01
Silver ion	-	-	0.003	-	-	-	-	-	-	0.008	0.008
Silver *	-	-	500	-	-	500	-	-	500	62.5	250

*Typical in inert carrier with 2% silver metal

Those actives that readily migrate from the plastic article to the immediate environment will yield a high concentration of biocide at the surface and remain in the vicinity for a period until the biocide is depleted. Migration may occur with the plasticiser and/or by a process of leaching, the latter being usually dependant on available moisture or oil and therefore the solubility of the active is of significance. In the case of silver the availability of sodium chloride may also contribute to the amount of surface available silver ions. Those plastics with low level or non-migratory biocides may not have sufficient active available on the surface to kill the microbes that may settle there.

Most bacteria and fungi will only grow and reproduce where water (liquid or vapour) is present. Some microorganisms will survive in the absence of water but will not grow, proliferate or cause deterioration. Laboratory microbiological testing can further assist in selecting the most appropriate active for the plastic substrate but this must be chosen on the basis of suitability for the application rather than suitability of the active ingredient being tested.

Thermal stability has long been a characteristic promoted by the inorganic biocide companies as a main advantage over organic biocides. However, processing temperature is a very one-dimensional assessment of organic biocide stability. It is also a question of processing time, acceptable loss of activity and whether any discolouration of the plastic is tolerable.

Artificial weathering is a tool to provide an indication of performance but cannot reproduce the natural weathering process realistically. The leach and UV effect of accelerated weathering and immediate microbial challenge test thereafter might affect the top layers of a PVC matrix, especially when testing thick films. It will not necessarily allow the migration of biocide from the bulk of a PVC material to the surface with slowly migrating biocides or may exaggerate effects on less UV stable actives near the surface. Accelerated tests portray these actives negatively whereas in practice they may perform well.

Photo 1: Effect of artificial weathering on test samples (ISO 846 B test method)

	Control	INTERCIDE ZNP (Zinc pyrithione)	INTERCIDE DBF 10 SVC (pellet type DCOIT)	INTERCIDE ABF 5 SVC (pellet type OBPA)
Addition levels		0.5%	1.2%	1.3%

Unleached & unweathered.

Leached.

QUV weathered.

Without leaching ZNP and OPBA show clear zones of inhibition, OBPA also after leaching. After QUV weathering all samples supported growth but DCOIT the least.

Shape or thickness of a PVC and effect on the antimicrobial performance

A thicker film of PVC will have a larger depot of biocide available compared to a thinner film. The lifetime of a thick film may be comparatively longer since the biocide in thicker films can migrate and replenish depleted active. Thin films are difficult to protect for any significant time period for this reason.

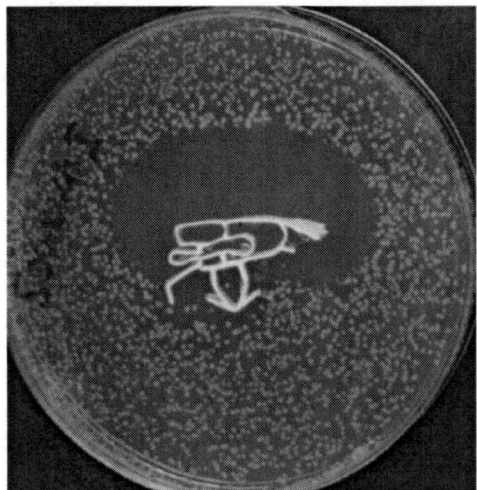

This depot effect can be demonstrated in photo 2 where the PVC profile does not produce a uniform zone of inhibition.

Photo 2: Refrigerator Seal profile containing biocide overlaid by agar containing *Staphylococcus aureus*

Artificial QUV and natural weathering processes will also differ depending on the thickness or shape of the PVC; accelerated weathering such as QUV or leaching cannot accurately mimic conditions in practice, partly because of the huge variability in geographical weather conditions. Although they may give an indication of performance, such artificial conditions can produce exaggerated UV and the hydroscopic effects. This may not be so pronounced or so intense in the field because a 'depot' of biocide may be present within a plastic matrix. The shifting of biocide within a PVC matrix may be the limiting factor rather than the hydroscopic effect of artificial weathering.

Figure 1: Proposed effects of QUV versus natural weathering on PVC films

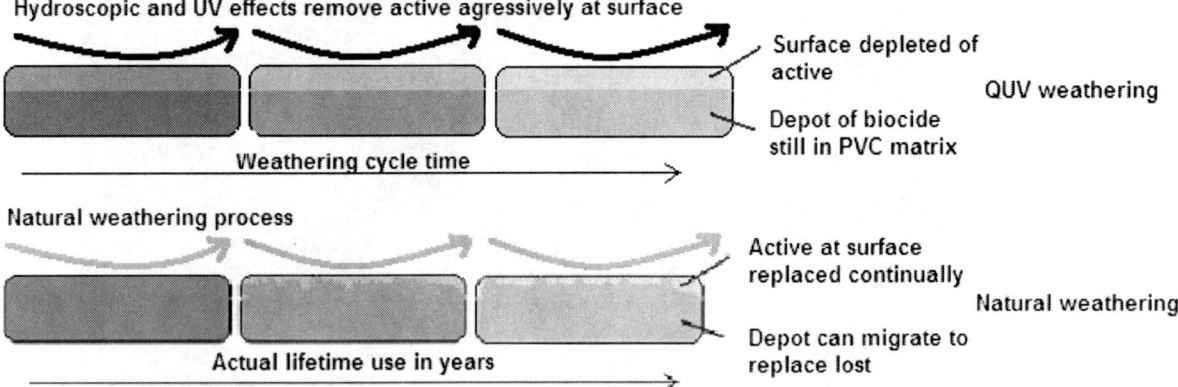

Function and effects of biocides in PVC

Various test methods are used by a laboratory to demonstrate the effect of the biocide in/on PVC. The intended function and desired effect must be clear to the test laboratory and the customer. Biocides in PVC are not capable of disinfecting a surface and some are only incorporated to protect the material itself rather than impart an external antimicrobial effect. Each biocide will have a specific mode of action and speed of effectiveness which will determine its suitability in certain scenarios.

Table 2: Comparison of effects

Biocide to protect material	Antimicrobial effect	Disinfectant
Incorporated into plastics	Within or coated on	Surface applied
Anti-fungi, algae and bacteria	Mostly antibacterial	Mostly antibacterial
Prevent discolouration	Prevent discolouration	Prevent discolouration
Stop cracking	Prevent odours	Prevent odours
Prevent pitting	Impart 'feel good factor'	Very rapid sterilising effect
Prevent brittleness	Improve product hygiene	Eliminate contamination
Improve product life span	Reduce contamination	Short life span typically
Maintain aesthetic appeal	Intended long life	
Medium to long life span		

PVC tends to be non-polar and hydrophobic in character, causing water droplets to run away or pool. 'Wetting' of the PVC surface may be enhanced by the presence of soiling or a surfactant. Agar plate tests attempt to mitigate the problem of surface wetting by creating a high humidity environment in a petri dish.

However it is probably not appropriate for PVC to be tested in such an intimate way which encourages migration of water soluble actives more readily than may otherwise occur. Artificial weathering (exposure to UV light, condensation and water spray) or leaching to PVC before microbiological tests might alleviate the hydrophobic effect.

Most laboratory tests are adopted in order to provide relatively rapid results. They serve as a model to mimic conditions found in practice, but many are not reliable in this respect, since they do not take account of factors such as: the degree of soiling, the potential level of contamination, presence of cleaning residues, increased surface area caused by abrasion, the age, exposure to weather and so on.

Agar plate techniques evaluate the growth of microorganisms on the test PVC and also in the surrounding agar. A zone of no growth (zone of inhibition) can either indicate an effective, a high concentration, a high rate of migration or leaching of the biocide active from the PVC into the surrounding media. However large zones of inhibition can indicate vulnerability to leaching. It is even possible for the plastic to become depleted of biocide and support microbial growth and yet still have a large zone of inhibition. Thus care must be taken when interpreting such test results. Most methods can be modified to suit applications, but for specific field use and to account for long term efficiency, methods such as humidity chamber, soil burial or vermiculite bed may be more suitable to provide more faithful conditions. Often a combination of tests is worthwhile in order to reflect performance under various climatic and environmental conditions but sometimes varying results can cause confusion.

Selected active ingredients

There are many organic and inorganic actives available for use in plasticised PVC and other plastics. The major actives are summarised below:

OBPA

Chemical name:
10,10 Oxybisphenoxarsine (OBPA)

Profile:
The INTERCIDE ABF range from Akcros Chemicals is typically available in 2% or 5% concentration in plasticiser or PVC pellets.
Since this biocide has a broad spectrum of toxicity the main concern on use of arsine is based on the potential leaching to the environment at the end of a PVC products lifecycle as well as the toxicity in handling concentrates.

Activity:
The levels of arsine in OBPA are fairly low but often sufficient to achieve the minimum inhibitory concentration (MIC) against the majority of fungi and bacteria of 10mg/litre. *Scopulariosis brevicaule (formerly classified Penicillium)* is known to produce arsenic based by-products and is therefore inherently resistant to arsenic but this rarely grows on PVC.

Field Performance:
Upsher & Roseblade have reported in 1984 that long-term activity of OPBA exposed at a jungle site is limited due to its leachability. Additionally results of Redlich, G., (c. 2000) highlight poor UV stability. Since it has a relatively low water solubility of 5 mg/litre this causes confusion. OBPA however has a tendency to migrate with some plasticisers used in PVC. It thus creates large zones of inhibition in some agar plate test methods.

Laboratory Performance:
Large zones of inhibition created by OBPA have historically been used as a marketing tool for OBPA suppliers to extol the advantages over other actives that may not produce such large zones in typical ASTM G21 or ISO 846 B test methods. However, potentially a large zone may mean that any active ingredients are lost fairly readily in comparison to actives which exhibit smaller zones in such tests.

Chemical Stability:
No major chemical stability issues.
Processing stability:
Generally very stable at most PVC processing temperatures.
Not very effective in the presence of tin stabilisers as the tin creates a layer through which the OBPA struggles to migrate and then be available at the surface to destroy microbes. May in some cases however prolong the lifetime of the OBPA conversely where a small quantity of tin is present.

OIT

Chemical Name:
N-Octyl-isothiazolin-3-one (OIT)

Profile:
The INTERCIDE OBF range from Akcros Chemicals is typically available in 8-20% concentration in plasticiser or PVA/PVC copolymer pellet. OIT is classed as a skin sensitiser in its liquid form, although once bound in PVC poses no hazard in most applications and is relatively biodegradable.

Activity:
This active is weak against some bacteria, therefore not widely used as a sole active for this purpose. It is an excellent fungicide, with minimum inhibitory concentration (MIC) of 1-10 mg/litre against most fungi and a good algicide but due to leachability in exterior applications has a short lifetime of activity in this respect. OIT works by interaction with cell amino acids affecting cell respiration and ATP synthesis (part of the Krebs cycle utilising sulphur containing amino acids) but this process is two-fold with initial 'inhibition' of the cell and then final biocidal action by halting cell repair mechanisms. OIT therefore has a relatively slow mode of action when compared to actives of general cell toxicity (such as OBPA) and those which disrupt cell walls. This is borne out by test methods requiring a rapid kill effect (such as ISO 22196).

Field Performance:
OIT has relatively high water solubility of 480mg/l at 25^0C which confers some advantages; providing good anti-fungal effects in interior applications. Conversely, it can be subject to excessive leaching where precipitation is high in exterior locations. So rarely used for exterior applications but shows robust protection for internal products and is one of the main replacements for heavy metal free alternatives to OPBA for interior applications.

Laboratory Performance:
OIT shows typically good results in unleached samples using ISO 846 and similar test methods. It is widely adopted for interior flooring and building materials and can provide some effect against bacteria in ISO 22196 methods, but is species dependant and requires relatively high addition levels compared to those required for fungi.

Chemical Stability:
OIT is inactivated by reducing agents (such as mercaptides and phosphites) and therefore this biocide should not be pre-mixed with metal stabilisers (as these often contain phosphites).

Processing stability:
OIT is heat stable up to 160^0C for 30 minutes and longer periods at lower temperatures. It is less stable at higher temperatures and some activity may be lost by reaction with reducing agents present in plastisols.

Figure 2: OIT% remaining over time following extrusion at specified temperatures

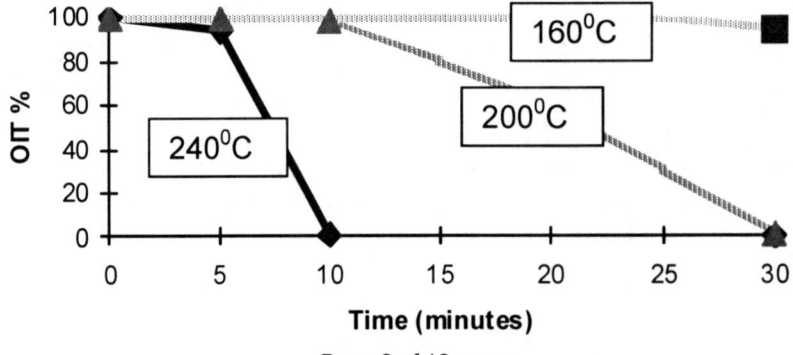

DCOIT

Chemical name:
4,5-Dichloro-n-octyl-isothiazolin-3-one

Profile:
The INTERCIDE DBF range from Akcros Chemicals is typically available as a 5 to 20% solution in plasticisers or EVA pellets. It is also a strong skin sensitiser in liquid form, although it is a waxy solid in neat form at room temperature.

Activity:
The double chlorinated octyl isothiazolinone has much lower solubility in water, 2mg/litre, compared to the non-chlorinated OIT. This makes it more suitable for exterior use, especially as it is also effective against algae. It confers both anti-fungal and some anti-bacterial effects with MIC varying from 5 ppm to 100 ppm against fungi and algae but it is weaker against *Trichoderma viride*.

Field Performance:
Because of DCOIT's low water solubility and good UV stability, it is ideal for external environments and for permanently or moisture prone products such as pool liners, awnings, tarpaulins and similar uses. It is widely used in these applications and furthermore once released to the environment it is readily degraded which makes it an environmentally favourable alternative to OBPA.

Laboratory performance:
For some applications where the DCOIT is not readily available on the PVC surface (interior, slightly damp applications) it can fail to perform adequately and fails some test protocols because of its very low migration levels. However, in field tests it consistently outperforms other actives that usually fare better in laboratory zone of inhibition type agar plate tests.

Chemical Stability:
Degradation occurs with some mercapto tin stabilisers and other reducing agents. Recovery of all the theoretical active by HPLC following processing is not possible. This was thought to be due to DCOIT becoming bound in the PVC matrix and therefore is unable to be fully extracted. This may be partly true but may also be due to chemical loss by interaction with the other elements within a PVC formulation.

Photo 3:
Colouration after 4 hours at 40^{0}C with DCOIT based actives mixed with Lankromark LZB 325 a barium zinc stabiliser (containing phosphite) DCOIT loss was also measured by HPLC.

Processing stability:
DCOIT is not quite as heat stable compared to other actives such as OIT (n-octyl isothiazolinone) but can remain stable at 160^{0}C for 30 minutes or more.

Zinc Pyrithione

Chemical Name;
2-Mercaptopyridine N-oxide zinc salt

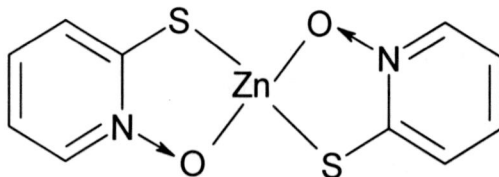

Profile:
The INTERCIDE ZNP range from Akcros Chemicals and the active is well known for its use as an anti-dandruff agent, it is typically supplied as a powder that has solubility in water of 20mg/litre. Some paste and pellet versions are also available for PVC.

Activity:
It is often promoted as an anti-bacterial additive although its activity is mainly anti-fungal. MICs against fungi range from only 7.5 ppm against *Alternaria alternata* to 50 ppm against species such as *Trichoderma* and up to 100 ppm for *Aspergillus niger*. The range is similar for bacteria, with only 10 ppm against *Staphylococcus sp.* but up to 400 ppm for *Pseudomonas aeruginosa*. Dose rates required to achieve good efficacy vary from 1000 ppm to more than 4000 ppm active ingredient depending on product and processing conditions.

Field Performance:
It is not widely used in PVC applications, except where the stabiliser levels are also controlled to ensure there is no excess Zinc but when used it is typically adopted for hospital and medical films.

Laboratory performance:
Zinc pyrithione shows good activity against fungi and algae and also performs as bacteriostat.

Photo 4: Zinc pyrithione activity against bacteria (E.coli above, *Staphylococcus aureus below* - Streak Plate based on AATCC Test Method 147-2004)

Control	Zinc Pyrithione	Zinc Pyrithione	OIT

Chemical Stability:
Zinc pyrithione in some PVC formulations, particularly those calendered or extruded under higher sheer, which are zinc sensitive can react with hydrogen chloride to form zinc chloride, which catalyses further degradation of the PVC. The outcome is yellowing and poor PVC stability.

Processing stability:
Zinc pyrithione contains typically 20% zinc. Patented systems (Burley/Clifford, 2003) using a hydrotalcite (magnesium and aluminium hydroxides) can be used with calcium/zinc heat stabilisers to scavenge chloride ions (acid scavenger) effectively swapping them for the carbonate ions within the layered lattice.

Folpet

Chemical Name:
N-Trichloromethylthiophthalimide

Profile:
The INTERCIDE FLPT range from Akcros Chemicals, it is a N-Haloalkylthio compound which is an off-white powder and is used in some plastics applications.
Folpet solubility in water is only 1 ppm with low solubility in solvents too, so any plasticiser based formulations are in paste form.

Activity:
Folpet is weak against some fungal species such as *Trichoderma sp.* (MIC 500ppm) and *Alternaria sp.* which means Folpet is rarely used in isolation or higher concentrations are required when used alone. Folpet also has some activity against bacteria but quite high levels may be required against organisms such as *Pseudomonas sp.* for example.

Field Performance:
It is not so widely used in Europe but its low water solubility coupled with a minimum inhibitory concentration against algae of 20ppm makes it ideal for exterior applications.

Laboratory performance:
Folpet is prone to yellowing in some applications, under QUV exposure especially, but at the appropriate dose gives good anti-fungal performance. Also it is claimed to be a bacteriostat but at relatively high levels.

Photo 5:
Samples of Oxime silicone sealant showing darkening of Folpet after QUV exposure (2 replicates) but effective anti-fungal performance at 0.25% a.i. or more.

A Standard (restricted use - R61)	B Blank	+ 0.5% INTERCIDE FLPT	+ 1.0% INTERCIDE FLPT	+ 0.5% FLPT 25 DINP (0.125% a.i.)	+ 1.0% FLPT 25 DINP (0.25% a.i.)

Chemical Stability:
Folpet is reactive with sulphides and mercapto compounds as with isothiazolinones.

Processing stability:
It is a heat stable molecule for most PVC applications, although decomposition starts at approximately 180°C.

Triclosan

Chemical Name: 2,2,4-dicholoro-2-hydroxydiphenyl ether

Profile:
This biocide is promoted by a number of companies and is not in the Akcros product portfolio.
Triclosan is widely used as an anti-bacterial additive in products such as toothpaste.
It is proposed that Triclosan acts as a site-specific inhibitor by mimicking a natural chemical enoyl-ACP, present in bacteria but not humans.

Activity:
Triclosan has a low minimum inhibitory concentration (MIC) of only 0.01ppm against *Staphylococcus sp.* but may not be so effective against other bacteria such as *Pseudomonas aeruginosa,* where the MIC is greater than 100ppm. The low rate of migration makes this biocide suitable for applications where this property is required, such as food contact articles.

Field Performance:
Proven worth in the field but under great scrutiny at present as may release dioxins (McNeill 2003).

Laboratory performance:
Triclosan has a low water solubility of 10mg/l at $20^{0}C$ which means that there is potentially poor availability of active at the surface in some applications, particularly dry interior uses.

Chemical Stability:
It is unstable under UV light and can cause colouration in contact with trace heavy metals (in alkaline conditions and after heating).

Processing stability:
Heat stable up to quite high temperatures ($275^{0}C$) so can be used in thermoplastic applications.

Silver

Chemical Name:
Silver ion, usually incorporated into an inert compound such as naturally occurring ceramic, zeolite or glass material.

Profile:
This biocide active is promoted by a number of companies but is not currently in the Akcros product portfolio. Silver use is prevalent for anti-bacterial applications. Silver is relatively non-hazardous and as FDA approvals for food contact applications (Markarian, 2002)

Activity:
Silver ions are said to be effective by interacting with multiple binding sites on the surface of the bacterial cell wall in addition to several other internal processes including interacting with sulfydryl groups (-SH) in proteins. The MIC data for silver seems to be significantly higher than the theoretical amount available at the surface of PVC but the bioavailability of silver is affected by the formation of various silver-halide complexes and may be influenced by sodium chloride in the microbiological medium used for susceptibility testing (Chopra 2007). It may similarly be unable to provide protection to 'soiled' articles as it will not be able to migrate sufficiently into the soiling or biofilm to provide protection.

Field Performance:
Silver is widely adopted for medical applications, it is claimed that moisture in the air causes low-level release that effectively maintains an antimicrobial surface. Silver is also used in food contact products.

Laboratory performance:
Silver performs very poorly in most agar plate test methods as the active does not migrate readily. Most methods demonstrating the effectiveness of silver utilise the ISO 22196 (or former Japanese Standard on which it is based) and similar intimate contact tests.

Chemical Stability:
Some PVC formulations, particularly those calendered or extruded under higher sheer, may be silver sensitive as silver can react with hydrogen chloride and form silver chloride, which catalyses further degradation of the PVC. The outcome can be severe darkening due to the silver chloride in a light source. The inert carrier of most silver chemistries permits a slower rate of release of silver ions and therefore minimise the discolouration associated with silver chemicals but this could also limit the bioavailability of the silver ion for efficacy requirements.

Processing stability:
The relatively low extrusion temperature of most PVC applications is not an issue for many organic actives. However, silver is tolerant to the high temperatures that are used in some extrusion processes.

Other actives

With such a variety of biocides available, it is perhaps useful to put a perspective on what are the main biocides of interest in the above list of active ingredients.

For antibacterial agents the use of Silver, Triclosan and Zinc pyrithione are prevalent. For use as an antifungal agent, traditional products such as OBPA are now being replaced largely by isothiazolinones. All products have some deficiencies or gaps in the spectrum of their activity and some mixed products are therefore offered. Mixed products combine the spectrum of activity and leachability of the different actives but can provide curious results in testing. In Photo 6 a combination biocide produces two distinct zones of inhibition. After QUV weathering, the effect of the first (and readily leachable biocide) disappears.

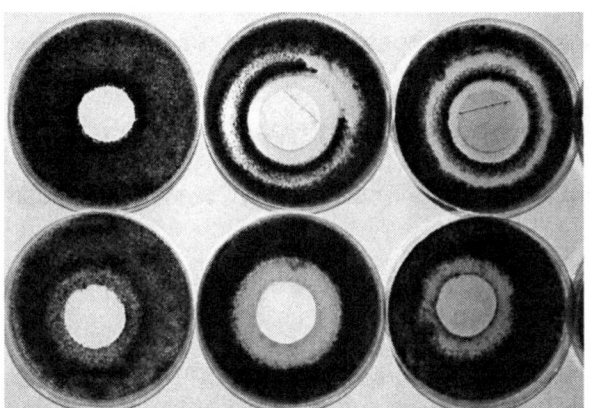

Photo 6:
Effect of combination biocides in ISO 846 B test without (above) and with (below) QUV weathering

Other actives of note include but are not limited to:

Zinc Oxide nanoparticles
The use of nanoparticles for 'self-clean' of a wide range of synthetic materials as well as PVC has been proposed and with added zinc metal imparts a similar activity to that proposed by silver. The large surface area offered by nano particles provides enhanced activity over standard zinc oxide.

IPBC (3-iodopropargyl-N-butylcarbamate)
INTERCIDE IPBC from Akcros Chemicals is supplied as a neat powder. It is an iodine containing fungicide used mainly in coatings but sometimes used in plastics applications. Its use in PVC at present is not widespread because it is not particularly stable at elevated temperatures and can be prone to strong discolouration/yellowing due to the iodine component. Degradation products can however still be microbiologically active and where discolouration is not an issue, as with backing layers and heavily pigmented PVC, IPBC can be used.

Butyl BIT (n-Butyl-1,2-benzisothiazolin-3-one)
This product has a similar solubility to the more commonly used octyl isothiazolinone and also supplied in liquid form. It gives similar performance to OIT. In comparable tests at extremes of temperature the stability of butyl BIT in a PVC formulation may not be as good as OIT however.

Summary

Every PVC formulation is different, so biocide suppliers and customers need a close relationship to ensure they are both aware of any potential compatibility and stability issues that might occur during processing in order to maximise biocide retention. Microbiological testing should be based on the end-use of the PVC. Appropriate test methods should be agreed to determine biocidal efficacy.

References and bibliography

ASTM G 21:1996. Determining the Resistance of Synthetic Polymeric Materials to Fungi.

ASTM E 1428:2009. Standard Test Method for Evaluation the Performance of Antimicrobials in or on Polymeric Solids Against Staining by *Streptoverticillium reticulum (A Pink Stain Organism)*

Burley J.W. and Clifford P.D., 2004. Extending the use of zinc-containing biocides in PVC, Journal of vinyl and additive technology, 10, No.2, June 2004, p95-98

ISO 16869: 2008. Plastics – Assessment of the Efficacy of Fungistatic Compounds in Plastic Formulations.

ISO 846:1997. Plastics - Evaluation of the Action of Microorganisms.

ISO 22196:2007. Plastics — Measurement of antibacterial activity on plastic surfaces

Ian Chopra, 2007. The increasing use of silver-based products as antimicrobial agents: a useful development or cause for concern, Journal of Antimicrobial Chemotherapy **59**, 587-590

McNeill, K. et al., 2003. Photochemical conversion of Triclosan to 2,8-dichlorodibenzo-p-dioxin in aqueous solution. J Photochem Photobiol A 2003, Article in Press.

Nichols, D. M., 2002. Antimicrobial Additives in Plastics and the European Biocidal Products Directive. Plastics Additives and Compounding. December, 2002.

D. Nichols (2004), Biocides in Plastics, RAPRA Review Reports Volume 15, No 12, 2004

Webb, J. et al., 2000. Fungal Colonization and Biodeterioration of Plasticized Polyvinyl Chloride. American Society for Microbiology. Vol. 66, No 8, August 2000.

Bibliography:

Wilfried Paulus, 1993. Microbiocides for the Protection of Materials - A Handbook, 1st Edition 1993. Chapman & Hall

THE USE OF REACTIVE SILANE CHEMISTRIES TO PROVIDE DURABLE, NON-LEACHING ANTIMICROBIAL SURFACES

Robert A. Monticello, PhD,
ÆGIS Environments
Midland, Michigan, 48640, USA
Tel: +1 989 832 8180 Fax: +1 989 832 7572 email: rmonticello@microbeshield.com

BIOGRAPHICAL NOTE

Robert A. Monticello received a B.S. in Microbiology from Michigan State University and a M.S. in Molecular Biology from Western Michigan University. He obtained his Ph.D. in Biochemistry and Molecular Medicine at the Wayne State University Medical School in Detroit, Michigan and completed a Post-Doctoral Fellowship at the Wayne State University Department of Molecular Medicine in Human Gene Therapy. After his formal educational training, he joined AEGIS Environments, a global manufacturer and supplier of antimicrobial agents, where he is currently a Vice President and is the Chief Technical Officer for the company.

ABSTRACT

The application of a Silane quaternary amine (Si-quat) based antimicrobial has been proven effective as a finishing agent on textiles and construction products for almost 40 years. Antimicrobial agents of this type have been used on a wide variety of porous and non-porous systems with outstanding results. Successful applications can be achieved using almost any type of wet process, such as a pad or spray and but may also be extruded or molded into various synthetic materials. Once the material is cured onto or into the substrate it can then provide the antimicrobial protection necessary to safeguard the product from microbial contamination and subsequent breakdown.

This paper and presentation will cover not only the ease of use of the Si-Quat antimicrobials but will provide a review of the key data and test techniques relating to the demonstration of efficacy, durability and utility in dealing with microbial problems on non-porous surfaces under real-world in-use conditions. Durability and real-life performance are critical factors when choosing the proper antimicrobial treatment. This eco-friendly product falls in line with the current emphasis on sustainability and environmental impact that is dominating the world markets.

INTRODUCTION

Almost all materials have one thing in common; they face a common enemy. Bacteria, fungi, algae, and other organisms can consume and degrade surfaces during shipment, storage, and use, causing loss of product as well as exposing the manufacturer to potential liability. Contamination and colonization of microorganisms on surfaces can result in problems as small as an offensive odor to serious human infections and death. Imparting an antimicrobial agent into synthetic material can create microbial resistant, non-porous surfaces that can alleviate many of these problems. However, selecting the right antimicrobial is essential to provide the appropriate protection to the product as well as to protect our environment. The list of available agents becomes limited when the criteria selection includes durability, regulatory approvals (EU BPD, US EPA), spectrum of activity, and toxicity to both the manufacturer and the end-user.

The use of a silane quaternary amine based antimicrobial can provide durable antimicrobial protection against a wide variety of microorganisms without the worry of leaching heavy metals, phenolic compounds or other toxic compounds that continue to contaminate our environment and present situations that promote microbial resistance.

Altering surfaces with durable non-leaching antimicrobial agents such that they provide an active killing "field" for killing one celled organisms on contact is a reasonable and attainable goal. The use of quaternized nitrogen silanes has been demonstrated to provide such treatments on a wide variety of surfaces and end-use conditions. There are many ways to modify surfaces so that they are less receptive to the settling, attachment, and colonization of microorganisms. These modifications can create surfaces so that microorganisms that come into direct contact with the treated surface are inhibited or killed or more easily cleaned away.

Chemical and physical bonding mechanisms using covalent bonding mechanisms, using covalent or ionic associations done by simple condensation reactions, energy induced as in plasma deposition, or catalyzed reactions of reactive materials have been demonstrated. The success of these surface modifications at controlling the deposition, attachment and propagation of good (useful) or bad (destructive, interfering, or annoying) microorganisms has often been limited by many factors. These factors include the lack of durability of the coating and the practical and cost effective application of these agents during product manufacturing. Such is the challenge to find technologies that can be evaluated and utilized in a safe, long lasting, and cost effective manner. Silane quat monomeric agents can both self crosslink and can link with available surface sites to create fully cured polymer that binds directly to the surface providing an antimicrobial coating that becomes part of the substrate itself. The non-leaching behavior of such a reactive surface allows for the control of surface microbial contamination without the continuous release of toxic components into the environment which can promote the formation of resistant organisms.

MICROORGANISMS

Mold, mildew, fungus, yeast, bacteria, and virus (microorganisms), are part of our everyday lives. There are both good and bad types of microorganisms. The thousands of species of microorganisms that exist are found everywhere in the environment, on our garments, on our bodies and on virtually every surface around us. Microorganisms, their body parts, metabolic products, and reproductive parts, cause multiple problems to synthetic materials. They are human irritants, sensitizers, toxic -response agents, causers of disease, and simple discomforting agents. Clearly, microorganisms are the most potent pollutants in our environment, on our clothes, and on our furnishings.

Microorganisms need moisture, appropriate temperatures, nutrients, and most of them need to be associated with a surface. Textiles, apparel, bathrooms, carpets, draperies, wall coverings, furniture, bedding and ceiling tiles create ideal habitats for microorganisms due to the high levels of humidity seen in these environments during common use. Nutrients utilized by microorganisms can be organic material, inorganic material, and/or living tissue. For example, bacteria play an important role as part of the body's microflora, and along with the skin, are shed continuously. Given acceptable growth conditions, they can multiply from one organism to more than one billion in just 18 hours. Over time, microorganisms can form highly complicated and durable microbial colonies that attach themselves to surfaces. These microbial biofilms are a prime concern in the medical industry and must be controlled before they form on the surface themselves.

Microorganisms cause problems with raw materials and processing chemicals, wet processes in mills, roll or bulk goods in storage, finished goods in storage and transport, and goods as they are used by the consumer. They are also an annoyance and aesthetic problem to architects, builders, and home owners. The economic impact of microbial contamination is significant and the consumer interests and demands for protection are at an all time high.

ANTIMICROBIALS

The term antimicrobial refers to a broad range of technologies that provide varying degrees of protection for both organic and synthetic products against microorganisms. Antimicrobials are very different in their chemical nature, mode of action, impact on people and the environment, in-plant-handling characteristics, durability on various substrates, costs, and how they interact with good and bad microorganisms. Antimicrobials are used in and on a variety of substrates to control bacteria, fungi, and algae. This control reduces or eliminates the problems of deterioration, staining, odors, and health concerns that they cause. Additionally, antimicrobial agents may prevent the loss of product during transport and can potentially reduce legal liability when microbial contamination occurs.

In the broad array of microorganisms there are certainly both good and bad types. Antimicrobial strategies for bad organisms must include ensuring that non-target organisms are not affected or that adaptation of microorganisms is not encouraged. For instance, antimicrobial agents applied to textiles must control all microorganisms on the textile without leaching into the environment and affecting the natural biological skin flora. In addition, as sublethal doses of antimicrobial agents may lead to adaptation. The antimicrobial agent should not lose effectiveness over time and cannot diminish in effective concentration.

Antimicrobial agents can be classified in two main types; leaching and non-leaching. Leaching antimicrobial agents are defined as agents that must come off the treated substrate in order to exert the antimicrobial properties. Any antimicrobial agent that must enter the cell to work is considered a leaching agent. Non-leaching agents are fixed to the treated surface (usually by covalent bonds) and subsequently do not need to leave this surface to provide antimicrobial action. As these agents are physically attached, there is generally no means for removal and therefore no means to diminish the overall strength. The need for new and safer antimicrobial technologies is obvious. These new agents must be safer to the end-use, the applicator, and also to the earth. Antimicrobial agents that do not leach from the original treatment site can provide for this protection.

But even non-leaching is not enough. Antimicrobial agents in general must have broad spectrum antimicrobial activity (equally effective against bacteria, fungi, and algae), have little to risk to the product or to the people applying the product, must easily fit current production systems, must be environmentally friendly, and must be compliant with all global biocidal regulations (U.S. EPA, EU BPD, REACH).

SILANE QUATERNARY AMMONIUM COMPOUNDS

In the mid-1960's, researchers discovered that antimicrobial organofunctional silanes could be chemically bound to receptive substrates by what were believed to be Si-O linkages. The method was described as orienting the organofunctional silane in such a way that hydrolysable groups on the silicon atom were hydrolyzed to silanols and the silanols formed chemical bonds with each other and the substrate. The resultant surface modification, when an antimicrobial moiety such as quaternary nitrogen was included, provided for the antimicrobial to be oriented away from the surface [1].

The attachment of this chemical to surfaces appears to involve two processes. First and most important is a very rapid process that coats the substrate with the cationic species one molecule deep. This is an ion exchange process by which the cation of the silane quaternary ammonium compound replaces protons from water on the surface. It has long been known that most surfaces in contact with water generate negative electrical charges at the interface between water and the surface. This mechanism is further supported by data generated with a radioactive silane quaternary ammonium compound. During the treatment, depletion of the radioactivity from solution was almost immediate by an amount corresponding to that sufficient to cover the surface one layer deep, even on surfaces which contain no functionality. Similar results are published for many organic quaternary ammonium compounds. The second process is unique to materials such as silane quaternary ammonium compounds that have silicon functionality enabling them to polymerize, after they have coated the surface, to become almost irremovable even on surfaces with which they cannot react covalently. Covalent bonding to the surface can also occur and through a series of heating and cooling steps, it is also possible to have intermolecular polymerization creating interpenetrating network in which the reactive silane forms anchors for additional polymer formation. Once hydrolyzed, the silanol groups become functionalized and are able to react with itself and available sites on the surface to form a dense polysiloxane network with an extremely high cationic charge density capable of destroying microbes.

ANTIMICROBIAL ACTIVITY OF SILANE QUAT ANTIMICROBIAL

This section summarizes the broad spectrum antimicrobial activity of the Si-Quat antimicrobial agent applied onto a variety of both porous and non-porous surfaces. The data represent over 35 years of experience and microbiological and chemical testing measuring the effectiveness of the Si-Quat antimicrobial agent after being applied onto surfaces such as furniture, carpets, wood and vinyl flooring, non-woven textiles (air filters), aquariums, etc. Surfaces treated with the Si-Quat technology have been shown to be resistant to the formation of biofilm. This resistance is due to two specific mechanisms which will be described below.

Since inception in the mid-1960's, the antimicrobial activity of the [3-(trimethoxysilyl) propyldimethyloctadecyl] ammonium chloride (Si-Quat) has been studied extensively on a variety of treated surfaces. The antimicrobial activity of solid surfaces treated with the Si-Quat agent was first described by Isquith et al[1] and later elaborated on by others, most notably, by Speier and Malek[2]. In their study, dose dependent antibacterial activity was demonstrated against both the Gram – *Escherichia coli* and the Gram + *Staphylococcus aureus* after treating a solid surface of clearly defined dimensions. The rate of kill and surface kinetics of these treated surfaces were further defined and demonstrated by Isquith and McCollum[3]. This work was followed by a companion study which measured the broad spectrum antimicrobial activity against a mixed fungal spore suspension (A*spergillus niger*, A*spergillus flavus*, A*spergillus versicolor*, *Penicillium funiculosum, Chaetomium globosum*). With the use of radioactive tracers, Isquith and McCollum demonstrated that "biological activity of the Si-Quat bonded to surfaces may offer a method of surface protection without addition of the chemical to the environment". Algicidal (*Chlorophyta, Cyanophyta* and *Chrysophyta*) activity of the Si-Quat applied to glass was demonstrated by Walters et al[4].

Further work demonstrates the ability to apply this material to a variety of substrates. This work includes surfaces from glass and aquariums to entire hospitals (Walters et al[4]., Lewbart et al[5]., and Kemper et al[6]). Kemper studied the microbial colonization of environmental surfaces in hospitals and the effectiveness of the

Si-Quat to control these organisms. This 30 month study measured persistent antimicrobial activity on surfaces treated with the Si-Quat agent. Isquith demonstrated antimicrobial activity on a variety of porous and non-porous surfaces. The Si-Quat antimicrobial agent was applied to surfaces as diverse as stone and ceramic, cotton and wool, vinyl and viscose, aluminum, stainless steel, wood, rubber, plastic, and Formica (Isquith et al[1]). These authors state that these surfaces "were found to exhibit durable antimicrobial activity when treated with Si-Quat, against a spectrum of microorganisms of medical and economic importance". Further independent testing confirms antimicrobial activity on air filters and fabrics treated and used directly in the hospitals settings.

The property of the Si-Quat AEM 5772 Antimicrobial that provides for the physical contact and rupturing of the cell membranes of single celled organisms revolves around the chemical structure of the monomer and subsequent final polymer. Contact with the oleophilic moieties of the long carbon chain and high cationic charge density exerted by the quaternized nitrogen of the polymer by the cell membranes of single celled organisms causes the physical rupture and inactivation of the membrane and the inhibition and death of the microbe.

This active ingredient monomer, when applied to surfaces and polymerized, provides a mode of antimicrobial activity that physically ruptures the cell membranes of microorganisms by ionic association (cell membranes carry a negative charge) and lipophilic attraction (the C18 associating with the lipoprotein of the membrane) causing disruption and lyses of the microbial cell. Speier and Malek[7] showed this lysis on treated nonwoven fabric surfaces through electron microscopic observations. The distortion of the overall cellular structure could be seen on both Gram + and Gram − bacteria on treated and untreated surfaces. The depletion of the cellular electrochemical potential across the membrane and release of cytoplasmic materials provides complete destruction of the microbe.

CONTROLLING BIOFILM DEVELOPMENT

Microbial contamination and subsequent biofilm formation is a major cause of infection, contamination, and product deterioration. Controlling or even removing the biofilm after its development is difficult. A useful strategy is to control biofilm formation before it starts. For the prevention of biofilm formation, control of both adherence and colonization of the microorganisms on the substrate surface is critical. One of the strategies to prevent biofilm formation is to modify the physiochemical properties of a surface in order to minimize or reduce the attraction of the surface to the microorganism thereby controlling adherence. Reducing the attraction simplistically can be done either by manipulating the ionic charge of the surface altering the electrostatic interface or changing the hydrophobic/hydrophilic properties through surface energy manipulations (or both) (Gottenbos et al[8]).

Controlling or minimizing the adhesion of microorganisms to the surface can be done using several techniques. Strategies used in the modification of surface characteristics range from altering the physical properties of the surface via mechanical abrasion to covalently attaching functional components to the surface (Marshell[9], MacKintosh[10], Bouloussa[11]). However, controlling the physical surface properties through water repellency does not appear to be enough to prevent biofilm formation. Bacteria can still adhere to highly hydrophobic surfaces.

Creating an active antimicrobial surface will destroy the adhering microorganisms, single celled organisms, thereby preventing further proliferation. Several groups have recently studied the ability to permanently create antimicrobial surfaces by covalently binding cationic polymers directly to surfaces (Kenawy[12], Huang[13], Lin[14], Kurt[15]).

The idea of creating active antimicrobial surfaces via the treatment with non-leaching quaternary amine compounds is certainly not new as presented above and using very similar approaches to the Si-Quat technology, these groups have created highly active antimicrobial surfaces. Using elaborate application techniques, long poly quaternary chains could be produced that create varied chain length polymers on surfaces with varying thickness. This work is summarized well in the review by Kenawy et at[12]. These groups demonstrated that a high cationic charge density and specific chain length polymerization were critical in the formation of permanent, non-leaching biocidal surfaces. In theory these long chain quaternary polymers are permanently fixed to the surface via covalent linkages but act directly on the cell membrane. This interaction is either through a physical association with the membrane via the long polymeric carbon chains and/or through direct ion exchange reactions with specific membrane components. The ion exchange theories in particular are interesting with the evidence that high surface charge density is directly related to killing efficiency. The killing efficiency and required charge density is dependent on organism, cellular components, surface charge of particular organisms and growth rate. (Murata[16], Kugler[17], Neu[18]).

It is critical, of course, that to use an antimicrobial agent in the prevention of biofilm formation, the agent must be broad spectrum and active against the particular biofilm causing organisms. Demonstration of the broad spectrum antimicrobial activity of surfaces treated with the Si-Quat antimicrobial agent can be found in the peer reviewed literature on a monthly basis. The Si-Quat technology, as reference above, is specific against all tested organisms typically responsible for biofilm formation.

Somewhat stimulated by the renewed understanding of the role of Si-Quat modified surfaces in the prevention of biofilm formation, several investigators renewed the investigation of the relationship between surfaces treated with the ÆGIS Si-Quat chemistry and the formation of microbial biofilm. The application of the ÆGIS Si-Quat onto surfaces structurally changes the surface. To further understand the relationship between water repellency and adsorption on surfaces treated with the Si-Quat, researchers from North Carolina State University, College of Textiles applied the Si-Quat technology directly onto polyester textiles and measured the water absorptive properties. This group demonstrated that the siloxane polymer that forms upon final hydrolysis and condensation of the silane monomer is directly related to time, temperature, and pH of treatment solution. Both hydrophilic and hydrophobic surfaces could be created depending on application procedure (Abo El Ola et al.[19]) while antibacterial activity of the surface remained intact. Saito et al[20]., from Hiroshima University, used treated silica particles to measure the relationship between the adherence of Oral Streptococci and surface hydrophobicity and Zeta potential. Gottenbos et al[8] from the University of Groningen demonstrated both *in vitro* and *in vivo* activity of Si-Quat treated silicone rubber used in the biological implants. As an expansion of this work from the same laboratory, Oosterhof, measured the inhibitory effects of positively charged coatings on the viability of yeasts and bacteria in mixed biofilm. Significant reduction in both adherence and colonization of organisms associated with tracheoesophageal shunt prosthetic biofilm (Oosterhof et al.[21]).

The Si-Quat technology when applied to surfaces affects both the adhesion properties of microorganisms due to increased hydrophobic properties of the long carbon chain fully polymerized and also directly destroys one celled organisms on contact through mechanisms described above. Nikawa et al[22] from Hiroshima University studied both the adhesion and colonization of mixed biofilm suspensions as a means to control biofilm formation on medical devices. This group demonstrated that commercially pure wrought titanium treated with the Si-Quat technology significantly reduced the adherence and colonization of both *Candida albicans* and *Streptococcus mutans,* even when the surface was coated with a proteinaceous layer like saliva or serum. Clearly this biofilm control mechanism was directly related to both the decreased adhesion due to the hydrophobicity created by the octadecyl alkyl chain and also due to the killing of the quaternary ammonium which killed initial adherent cells and also retarded or inhibited subsequent microbial growth. Furthermore, cell culture and cytotoxicity studies were performed in order to demonstrated the non-leaching behavior of the antimicrobial coating. No significant cytotoxicity of Si-Quat was observed in cell viability tests or inflammatory assays.

SUMMARY AND CONCLUSIONS

The use of reactive silanes functionalized with antimicrobial agents has been demonstrated to provide surfaces which are resistant to microbial growth and subsequent biofilm formation. These surfaces become resistant due to both the biostatic repulsions of microorganisms to the surface and due to the highly charged cationic density and physical attraction of the resulting polymer network. These non-leaching antimicrobial surfaces can be applied to a variety of substrates due to the highly reactive silanol groups associated with the antimicrobial agent. These reactive groups bind both to the surface and itself forming highly cross-linked networks that form durable protective coatings on virtually any surface.

With the increase in awareness of multiple antibiotic resistant bacteria, the recognition of increased sensitivity of our environment that bioaccumulates toxic chemicals and formation of strict regulatory agencies, it is paramount that new uses for older, safer, antimicrobial agents are investigated. These antimicrobial agents must be safe for the environment and end-user but still protect our products from the detrimental affects caused my rampant microbial contamination. The use of reactive silane chemistry to provide durable, non-leaching antimicrobials on synthetic material has been demonstrated to be a way of controlling microbial contamination in a safe and effective manner

LITERATURE CITED

1   Isquith, A.J. et al, Applied Microbiology (Dec., 1972), p. 859-863.

2   Malek, J.R. and Speier, J.L., *Development of an Organosilicone Antimicrobial Agent for the Treatment of Surfaces*, J. for Coasted Fabrics, Vol. 12 (July, 1982), P. 38-46.

3   Isquith and McCollum. *Surface Kinetic Test Method for Determining Rate of Kill by an Antimicrobial Solid.* Applied and Environmental Microbiology, p. 700-704. November 1978.

4   Walters, P.A. et al, Applied Microbiology, Vol. 25, No. 2, (Feb., 1973), p. 256.

5   Lewbart et al. *Safety and Efficacy of the Environmental Products Group Masterflow Aquarium Management System with AEGIS Microbe Shield.* Aquacultural Engineering 19 (1999) p. 93-98.

6   Kemper et al. *Improved Control of Microbial Exposure Hazards in Hospitals: A 30-month Field study.* National convention for Association of Practitioner for Infection Control (APIC) .1992. Gottenbos et al, , *Antimicrobial Effects of Positively Charged Surfaces on Adhering Gram-positive and Gram-negative Bacteria*, Journal of Antimicrobial Chemotherapy (2001) 48, pp. 7-13.

7   Speier, J.L. and Malek, J.R., *Destruction of Microorganisms by Contact with Solid Surfaces*, J. of Colloid and Interface Science, Bol. 89, No. 1 (Sept. 1982), p. 68-76.

8   Gottenbos et al. *In vitro and in vivo Antimicrobial Activity of Covalently Coupled Quaternary Ammonium Silane Coatings on Silicone Rubber.* Biomaterials 2002; 23: 1417-1423.

9   Marshall et al. , J. Gen. Microbiology. 68 (1971), p. 337.

10 MacKintosh et al., *Effects of Biomaterial Surface Chemistry on the Adhesion and Biofilm Formation of Staphylococcus Epidermidis in Vitro,* Wiley Inter-Science (2006), pp. 836-842.

11 Bouloussa et al., *A New, Simple Approach to Confer Permanent Antimicrobial Properties to Hydroxylated Surfaces by Surface Functionalization*, Chem. Commun., 2008, pp. 951-953.

12 Kenawy et al, *The Chemistry and Applications of Antimicrobial Polymers: A State-of-the-*Roth, C., Canadian Pat. No. 2,010,782 (May 24, 1977).

13 Huang et al, *Nonleaching Antibacterial Glass Surfaces via "Grafting Onto": The Effect of the Number of Quaternary Ammonium Groups on Biocidal Activity*, Langmuir 2008, 24, pp.6785-6795.

14 Lin et al, *Insignts into Bactericidal Action of Surface-attached Poly (vinyl-N-hexylpyridinium) Chains*, Biotechnology Letters 24:2002, pp. 801-805.

15 Kurt et al, *Highly Effective Contact Antimicrobial Surfaces via Polymer Surface Modifiers*, Langmuir 2007, 23, pp. 4719-4723.

16 Murata et al. L, *Permanent, Non-leaching Antibacterial Surfaces-2: How High Density Cationic Surfaces Kill Bacterial Cells*, Biomaterials 28 (2007) pp. 4870-4879.

17 Kugler et al. *Evidence of a charge-density threshold for Optimum Efficiency of Biocidal Cationic Surfaces*, Microbiology (2005) 151, pp. 1341-1348.

18 Neu, T.R. *Significance of Bacterial Surface-Active Compounds in Interaction of Bacteria with Interfaces*, Microbiological Reviews, Mar. 1996, pp. 151-166.

19 Abo El Ola et al., *Unusual Polymerization of 3-(trimethoxysilyl)-propyldimethyloctadecyl ammonium chloride on PET Substrates*, Polymer 45 (2004) pp. 3215-3225.

20 Saito et al. *Adherence of Oral Streptococci to an Immobilized Antimicrobial Agent*, Archs Oral Biol. Vol. 42, No. 8, 1997 pp. 539-545.

21 Oosterhof et al., *Effects of Quaternary Ammonium Silane Coatings on Mixed Fungal and Bacterial Biofilms on Tracheoesophageal Shunt Prostheses*, Applied and Environmental Microbiology, May 2006, pp. 3673-3677.

22 Nikawa et al. *Immobilization of Octadecyl Ammonium Chloride on the Surface of Titanium and Its Effect on Microbial Colonization In Vitro.* Dental Materials Journal 24(4), 2005, pp.570-582.

ADDITIONAL REFERENCE LITERATURE

23 Gettings, R.L. and Triplett, B.L., *A New Durable Antimicrobial Finish for Textiles*, Book of Papers, AATCC National Conference (1978).

24 Hayes, S.F. and White, W.C., *How Antimicrobial Treatment can Improve Nonwovens*, American Dyestuff Reporter (June, 1984).

25 McGee, J.B. et al., *New Antimicrobial Treatment for Carpet Applications*, American Dyestuff Reporter, (June, 1983).

26 Lawrence, C.A., *Germicidal Properties of Cationic Surfactants*, Chap. 14, Cationic Surfactants

27 Baier, R.E., *Substrate Influences on Adhesion of Microorganisms and Their Resultant New Surface Properties*, Adsorption of Microorganisms to Surfaces (John Wiley and Sons, 1980).

28 Andresen et al., *Nonleaching Antimicrobial Films Prepared from Surface-Modified Microfibrillated Cellulose*, Biomacromolecules 2007, 8, pp. 2149-2155.

Lightning Source UK Ltd.
Milton Keynes UK
03 September 2010

159400UK00001B/8/P

9 781847 355546